PLANT DIVERSITY
OF AN ANDEAN CLOUD FOREST

Plant Diversity of an Andean Cloud Forest

Checklist of the Vascular Flora of Maquipucuna, Ecuador

Grady L. Webster and
Robert M. Rhode

UNIVERSITY OF CALIFORNIA PRESS
Berkeley • Los Angeles • London

Volume 82

UNIVERSITY OF CALIFORNIA PRESS
BERKELEY AND LOS ANGELES, CALIFORNIA

UNIVERSITY OF CALIFORNIA PRESS, LTD.
LONDON, ENGLAND

Library of Congress Cataloging-in-Publication Data

Webster, Grady L. (Grady Linder), 1927–
    Plant diversity of an Andean cloud forest : inventory of the vascular flora
of Maquipucuna, Ecuador / Grady L. Webster and Robert M. Rhode.
        p. cm. — (University of California publications in botany ; v. 82)
    Includes bibliographical references (p.)
    ISBN 0-520-09830-7 (pbk. : alk. paper)
    1. Botany—Ecuador—Bosque Protector Maquipucuna Region.
2. Bosque Protector Maquipucuna (Ecuador) I. Rhode, Robert M. (Robert
Martin), 1961– II. Title III. Series.

QK266.W43 2001
581.9866'13—dc21   00-053221

The paper used in this publication meets the minimum requirements
of ANSI/NISO Z39.48-1992 (R 1997) (Permanence of Paper)

Dedicated to

Padre Luis Sodiro, S. J.

(1836–1909)

Pioneer plant systematist in Ecuador

and

Botanical explorer of the land of the Yumbos

Contents

Illustrations

Figures

Plates

Characteristic Maquipucuna monocots

Tables

Acknowledgments

The completion of this project has involved a decade of effort by a number of individuals working at different institutions. The plants enumerated in this checklist have mostly been collected during a ten year period, from 1989 through 1998, by Grady L. Webster and collaborators on trips sponsored by the University of California Research Expeditions Program (UREP). During this time, we received the indispensable support of Ellen Dean, Director of the University of California (Davis) Herbarium, who has assisted in many ways throughout the project, including the establishment of a reference herbarium collection of the Maquipucuna flora. She also reviewed our treatment of the Solanaceae and determined a number of collections. We are especially indebted to Jean Colvin, Director of UREP, who took an active interest in our work and helped provide the essential logistic support. The success of our project owes a great deal to Rodrigo Ontoneda and Rebeca Justicia, founders and directors of the Fundación Maquipucuna, who created and sustained a supportive environment for our researches. The staff at Fundación Maquipucuna, especially including Abi Rome and Mike Dilger, provided helpful logistical support.

An important role was played by graduate students from the University of California (Davis), who acted as field assistants: Piero Delprete and Roberto Urtecho (1989); Michael McComb (1990); Dean Kelch (1990, 1991, & 1996); Fred Hrusa (1992); Robert Rhode (1992 & 1995); and Brian Smith (1993). Ecuadorian graduate students who participated include Montserrat Rios (1989, 1990) and Daniela Andrade (1992).

We are particularly indebted to Sr. Bernardo Castro, forest officer for the Fundación Maquipucuna, sometimes assisted by Luiz Pozo; Bernardo Castro's generous and skilled assistance has been critical in securing material of rare and inaccessible plants. Collections of Maquipucuna plants have also been provided by a number of botanists on brief visits, including Carlos Cerón, Carlos Quelal, and Galo Tipaz, who have collected voucher specimens at a permanent ecological plot. On two trips in 1990 and 1991, Alwyn Gentry collected transect vouchers (with B. Boyle, D. Rubio, & R. Valencia). In 1996, a significant collection of ferns was made by Dr. Kenneth A. Wilson. Assistance was also provided by other researchers at Maquipucuna, including Robert Raguso, Durrel Kapan, and Zak

Zahawi. A number of others have made short collecting trips to the reserve, including Henrik Balslev, Robbin Moran, David Neill, Walter Palacios, and Larry Skog.

We are indebted for assistance to David Neill at the Herbario Nacional in Quito (QCNE) for many courtesies, including but not limited to: providing facilities for drying specimens, helping with permits for collecting, and shipping specimens. Other botanists in Quito also gave assistance, including Carlos Cerón (Universidad Central), Efrain Freire (QCNE), and Benjamin Øllgaard (University of Aarhus and the Herbario, Pontificia Universidad Católica). Montserrat Rios (Pontificia Universidad Católica) gave us valuable support in organizing field work in the early stages of the project, and in providing vernacular names. The staff at the Missouri Botanical Garden provided determinations and other assistance, especially a critical review of the first draft of the manuscript by Peter Jørgensen. Extremely important assistance has been provided by the specialists who have generously contributed determinations of the UREP collections. Several specialists made very significant contributions to the checklist. Alan Smith (UC) reviewed the entire section on pteridophytes and most of the determinations. Harold Robinson furnished a large number of determinations of Asteraceae and described a new species of *Clibadium*. Piero Delprete (NY) reviewed the treatment of the Rubiaceae, as well as providing his own collections and determinations. Ricardo Callejas (HUA) furnished many determinations for the difficult family Piperaceae. Henk van der Werff not only determined specimens of the critical family Lauraceae, but also made significant collections in the Alambi Valley. James Luteyn, Larry Skog and Caloway Dodson determined most of the collections of the species-rich families Ericaceae, Gesneriaceae, and Orchidaceae, respectively.

Other major contributors of determinations include: Frank Almeda (Melastomataceae), Henrik Balslev (Arecaceae), J. Beckner (Orchidaceae), C. C. Berg (Cecropiaceae and *Ficus*), Paul Berry (Onagraceae) Allen Bornstein (Piperaceae), K. Camelbeke (Cyperaceae), Carlos Cerón (various taxa), E. Cotton (Melastomataceae), Theodore S. Cochrane (Capparidaceae), Gerritt Davidse (Poaceae), Bente Eriksen (*Monnina*), R. Erikson (Cyclanthaceae), Robert Faden (Commelinaceae), Barry Hammel (Clusiaceae), Bruce Holst (Myrtaceae), Peter Jørgensen (Passifloraceae), Denis Kearns (Cucurbitaceae), Helen Kennedy (Marantaceae), Sandra Knapp (Solanaceae), W. J. Kress (Heliconiaceae), J. Kuijt (Loranthaceae, *s. lat.*), S. Laegaard (Poaceae), Ronald Liesner (Dichapetalaceae), J. A. Lombardi (Vitaceae), Harry Luther (Bromeliaceae), P. J. M. Maas (Zingiberaceae), M. Manazares (Bromeliaceae), Elizabeth McClintock (Hydrangeaceae), Lucinda McDade (Acanthaceae), John Mickel (*Elaphoglossum*), L. E. Mora-Oseja (Gunneraceae), Robbin Moran (ferns), Michael Nee (Solanaceae), David Neill (Fabaceae, *s. lat.*), Magnus Neuendorf (*Bomarea*), Benjamin Øllgaard (Lycopodiaceae), Walter Palacios (various taxa), John Pipoly (Myrsinaceae), Susanne Renner (*Siparuna*), Tony Reznicek (Cyperaceae),

Katherine Romolereux (Rosaceae), Lyman Smith and Stephen Smith (Begoniaceae), M. A. Spencer (Bromeliaceae), John Strother (Asteraceae), Charlotte Taylor (Rubiaceae), Carol Todzia (Chloranthaceae), Carmen Ulloa (Zingiberaceae), Ivan Valdespino (Selaginellaceae), A. L. Weitzman (Theaceae), Kenneth A. Wilson (ferns), and John Wurdack (Melastomataceae).

Special thanks are due to the UREP volunteers, without whose hard work the extensive collections underpinning this study could not have been assembled. The dates, numbers collected, and names of UREP participants for each of the seven trips are as follows:

1989 (26 Aug.–16 Sept.; 27001–27773): Susan Addison, Kevin Bainard, Owen Bubar, Alina Dorian, Diane Dzilvelis, Gregoire Goodstein, Lorraine Hebert, Patricia Horton, Molly O'Malley, Virginia Revotskie, Dekalb Russell, Robert Schilling, Naomi Shibata, Sharon Ungerleider, Therese Thompson, and Vicki Valverde (also with Piero Delprete, Montserrat Rios, and Roberto Urtecho).

1990 (7–25 July; 27775–28399): Robert Bonning, Earth Clemons, Anthony Cody, Gabrielle De Benedictis, Richard Koeman, Jamie Natelson, Raquel Ornelas, Therese Thompson, and Christine Vargas (also with Dean Kelch, Michael McComb).

1991 (2–12 July; 28675–28981): Alisa Barnhart, Roxanne Bittman, Sharon Christman, Steve Cotter, Jaqueline Elbing-Omania, Niall McCarten, Molly O'Malley, Phyllis Schmitt, Linda Sciaroni, and Phoebe Tanner (also with Dean Kelch).

1992 (6–16 July; 29032–29555): Louise Armstrong, Elfego Gomez, Leslie Hinton, Sandy Koshari, Carol & George Kuehn, Maria Laxo, Judy McEntyre, Hal McGrath, Pat Murray, Sally Scholl, and Cynthia Zurinsky (also with Daniela Andrade, Piero Delprete, Fred Hrusa, and Robert Rhode).

1993 (30 Aug.–9 Sept.; 29977–30617): Sigward Elsas, Eric Huang, David Isbister, Joanne Miller, Marie Morgan, Tom Paradise, and Amy Vanacore (also with Juan Carlos, Rodney Friend, and Brian Smith).

1995 (5–16 Jan.; 31000–31349): Tom Colby, Jim Cooper, Nancy James, Patricia Ledlie, Carol Lerner, Mitzi Schad, and John Stone (also with Robert Rhode).

1996 (12–28 June; 31501–32021): Erik Aguilar, Ron Bjork, Michael Canfield, Richard Hines, Alice Hunter, Pat Ledlie, Charleen Reuscher, and Kenneth A. Wilson (also with Dean Kelch, Bernardo Castro, and Alfonso del Hierro); a visiting botanist at Maquipucuna, Zak Zahawi, also provided collection data from the Hacienda del Carmen area.

More recently, two additional trips were also made to the reserve, and appreciation is extended to the participants:

1997 (25–29 June; 32330–32441): Grady Webster and Bernardo Castro.

1998 (10–20 November; 32731–32974): Grady Webster, Bernard Castro, Dean Kelch, Arcenio Barrera, and Alfonso del Hierro.

Thanks also go to Michael A. Vincent and R. James Hickey for their help and forbearance, as well as to the Biology Department of Rockhurst University.

Finally, we wish to thank our wives, Barbara Webster and Barbara Németh Rhode, for surpassing patience and support.

Abstract

This inventory summarizes the vascular flora of the Bosque Protector Maquipucuna and adjacent areas, in Cantón Quito, Provincia Pichincha, Ecuador. Within the boundaries of the floristic region (approximately 00°00' to 00°10' N and 78°35' to 78°41' W), the total area is about 22,000 ha, of which about 4,500 ha lie within the Maquipucuna reserve. The steep and dissected slopes of most of the area, lying between elevations 1100 m and 2800 m, are covered with lower montane and upper montane Andean cloud forest which receives over 3000 mm of precipitation per year.

The total vascular flora recorded in this checklist, based on about 5,200 collections, includes 155 families, 621 genera, and approximately 1,640 species, including 228 species of pteridophytes. The flora is rich in epiphytic taxa, especially Araceae, Bromeliaceae, Orchidaceae, Ericaceae, Gesneriaceae, and many ferns. The number of clearly exotic species (44) is very low. In its composition, the Maquipucuna flora resembles that of the Chocó area in Colombia, and it may be regarded as representing the southern terminus of the Chocó flora on the western slopes of the Ecuadorian Andes.

Resumen

Este inventorio presenta un resumen de las plantas vasculares que se encuentran en el Bosque Protector Maquipucuna y sus alrededores, en Canton Quito, Provincia Pichincha, Ecuador. Dentro de los límites de la región florística (aproximadamente 00°00' a 00°10' N y 78°35' a 78°41' O), el area total es de aproximadamente 22.000 ha, de las cuales unas 4.500 ha corresponden a la reserva Maquipucuna. Las laderas escarpadas y accidentadas entre 1100 y 2800 m están cubiertas con bosque lluvioso montano bajo y alto, los cuales reciben más de 3.000 mm de precipitación anual.

La flora vascular citada en éste inventorio, se basa en aproximadamente 5.200 colecciones, incluye 155 familias, 621 géneros, y 1.640 especies (incluyendo 228 especies de pteridófitas). La flora es abundante en plantas epifíticas, en particular las Araceas, Bromeliaceas, Orchidaceas, Ericaceas, Gesneriaceas, y numerosos helechos. El número de especies claramente exóticas (44) es muy bajo. En su composición la flora de Maquipucuna es similar a la flora de Chocó (Colombia) y se le puede considerar como el representante austral de la flora chocoense en las laderas occidentales de los Andes ecuatorianos.

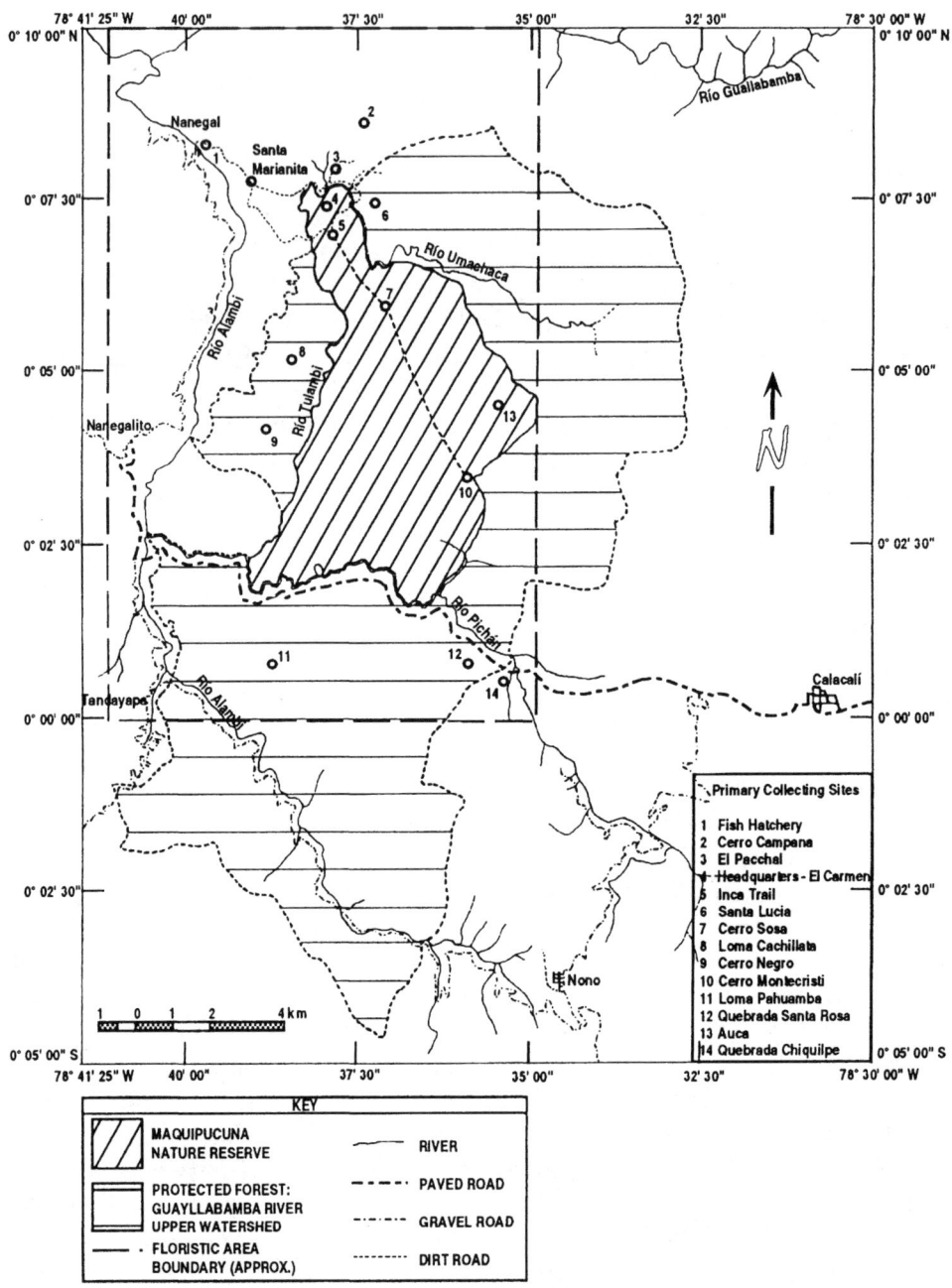

Primary Collecting Sites	
1	Fish Hatchery
2	Cerro Campana
3	El Pacchal
4	Headquarters - El Carmen
5	Inca Trail
6	Santa Lucia
7	Cerro Sosa
8	Loma Cachillata
9	Cerro Negro
10	Cerro Montecristi
11	Loma Pahuamba
12	Quebrada Santa Rosa
13	Auca
14	Quebrada Chiquilpe

KEY

MAQUIPUCUNA NATURE RESERVE

PROTECTED FOREST: GUAYLLABAMBA RIVER UPPER WATERSHED

FLORISTIC AREA BOUNDARY (APPROX.)

RIVER

PAVED ROAD

GRAVEL ROAD

DIRT ROAD

Figure 1. Map of the Maquipucuna region indicating major collecting areas; boundaries shown for Maquipucuna Nature Reserve do not reflect recently acquired areas. Adapted from base map courtesy of the Fundación Maquipucuna.

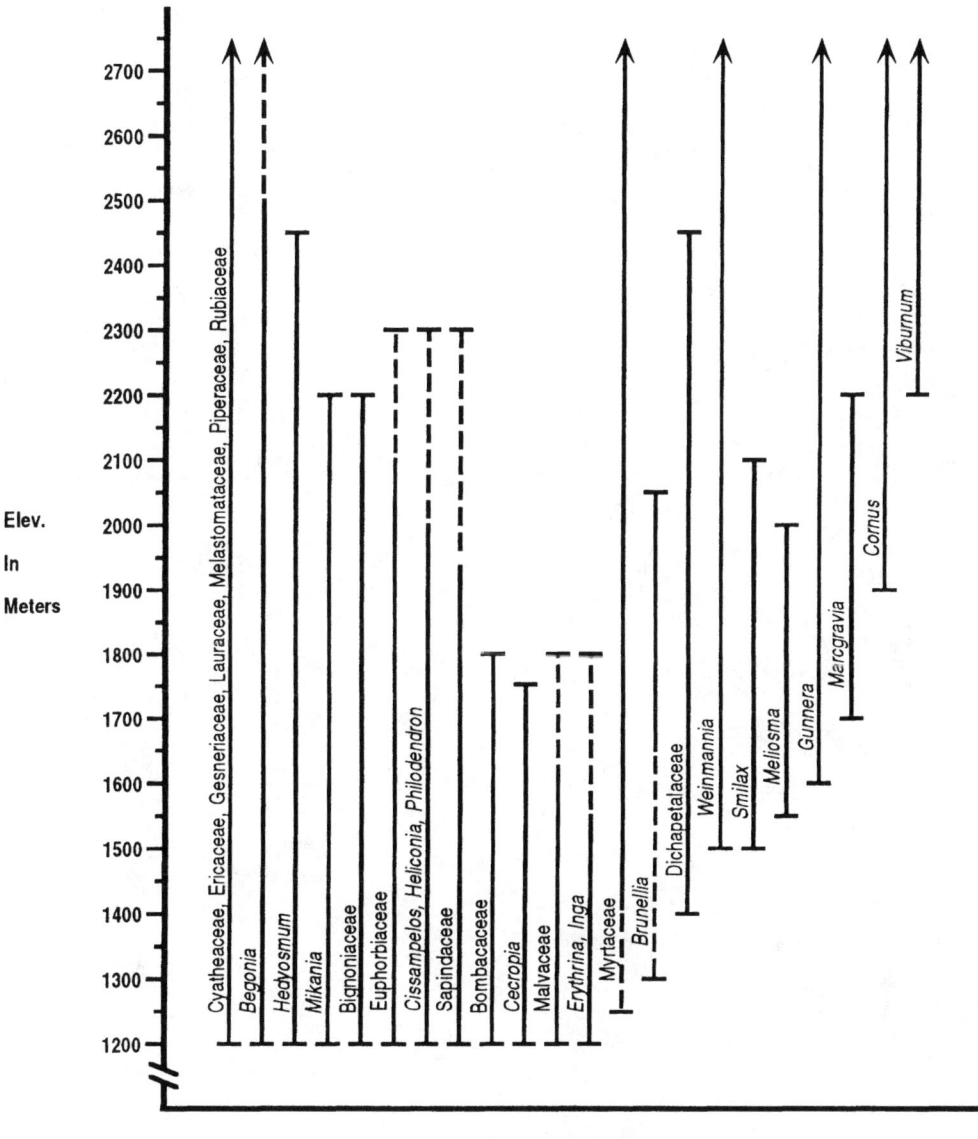

Figure 2. Elevational ranges of selected taxa at Maquipucuna, with lower and upper limits set at 1150/1200 and 2750 m; dashed portions of lines indicate extensions of elevation due to single species. The ranges of a number of taxa may extend higher elsewhere in Ecuador.

INTRODUCTION

The compelling necessity for inventories of neotropical plant diversity has recently been emphasized by Neill and Øllgaard (1993), and by Phillips and Raven (1996), who propose a strategy for sampling this diversity. This will evidently be possible only through creating the necessary organizational infrastructures and professional networks. At Maquipucuna, one of the sample sites listed by Phillips and Raven, two of the three inventory techniques—permanent plots (1 ha) and rapid assessment floristic samples (0.1 ha)—have been put in place, and the rapid assessment plot data of Gentry for Maquipucuna have been assembled (Phillips & Miller, 1996, ined.?). The checklist published here represents the application of the third technique, that of the floristic inventory of a limited region (1–100 km^2); such an inventory may serve as the precursor of a *florula* with keys and descriptions.

The Bosque Protector Maquipucuna, a private biological reserve of the Fundación Maquipucuna, lies on the western slopes of the Andes bounded by steep valleys southeast of the town of Nanegal, about 40 km northwest of Quito, Ecuador (Fig. 1); it is one of nine areas in Ecuador designated as Bosque Protector that are cited by Mena Vásconez (1995), but according to Rebeca Justicia (pers. com.) there are 77 such areas. The Maquipucuna reserve is very near the equator, extending latitudinally between 00°02′N and 00°08′N, and longitudinally between 78°35′ and 38′W. It includes at present about 4,500 hectares of mountainous terrain, increasing in elevation on the slopes of Cerro Sosa (Montañas de Maquipucuna) from c. 1200 m to nearly 2800 m at the summit of Cerro Montecristi, the highest point. Most of the reserve is covered by Andean cloud forest that has been greatly disturbed below 1500 m, but is largely pristine above. The boundaries used for the checklist, between the equator and 00°10′N and between 78°35′ and 41′W, include part of the buffer zone (upper Guayllabamba River basin) around the reserve.

Since its founding in 1988, the Maquipucuna reserve has been utilized for biological research by a number of investigators, but so far little of its biodiversity has been documented in print. Unpublished lists of bird species have been compiled (Marín et al., 1992; Greenfield, 1993), and a checklist of the butterflies has recently been published (Raguso, 1996). Svenning and Balslev (1998) have

1

provided an illustrated floristic treatment of the Maquipucuna palms. There is a recent study of plant succession in lower montane areas (Zahawi & Augspurger, 1999), and through the courtesy of Robin Foster (Field Museum, Chicago) a laminated sheet of photographs of 40 characteristic Maquipucuna species (Foster et al., 1999) has been issued. Currently, a long-term project to study the "below-ground ecology" of the rainforests at Maquipucuna is underway through collaboration between the Fundación Maquipucuna, the Institute of Ecology at the University of Georgia, and the University of Chicago (Sarmiento, 1995). It seems apparent that Maquipucuna is going to become an increasingly important site for ecological and systematic studies of Andean montane forests.

Beginning in 1989, botanists from the Herbarium at the University of California, Davis (DAV), under the auspices of the University of California Research Expeditions Program (UREP), have made annual trips to Maquipucuna in collaboration with the Fundación Maquipucuna, the Herbario Nacional (QCNE), and Herbario of the Pontificia Universidad Católica (QCA) in Quito. This checklist of the vascular flora of Maquipucuna is now presented with the hope that it will increase knowledge of Andean plant biogeography, facilitate ecological studies, and expedite conservation efforts to preserve the great wealth of biodiversity in Ecuadorian cloud forests. In accordance with the suggestions of Neill and Øllgaard (1993), this checklist is intended to be the precursor of an illustrated manual for identification of the vascular plants of Maquipucuna and adjacent cloud forest areas of the western Andean slopes in Ecuador.

CLIMATE AND TOPOGRAPHY

The climate of the Maquipucuna region is equatorial and wet, but unfortunately there are no long-term temperature or precipitation data from within the reserve. The closest weather station cited by Terán (1984), in Nanegalito, is indicated as having an annual rainfall of 3230 mm at an elevation of 1630 m. Sarmiento (1994) cites annual precipitation from Nanegal as 3198 mm; presumably the ridges and slopes above 2000 m have considerably greater precipitation. In the classification of Blandín Landívar (1977), the climate of the reserve should be "subtropical lluvioso" (more than 2000 mm of rainfall per year), with a mean annual temperature of less than 20° C. Because of the considerable elevational range in the reserve, the mean annual temperature probably varies from about 18° C near 1200 m to c. 10° C near the top of Cerro Montecristi. Despite the equatorial position of the reserve, there is a distinct dry season from June to September. Sarmiento (1994) has noted that Maquipucuna lies at the more humid western end of an elevational and precipitation gradient leading eastward to the drier slopes of the Pululahua Crater area in the Guayllabamba River basin. The transition zone to a drier climate begins approximately at the eastern boundary of the floristic area (78°35′W), where there are the westernmost outposts of a few relatively xeric ("ustic" in Sarmiento's terminology) taxa such as *Cheilanthes*, *Pellaea*, and *Echeveria*.

The topography of the Cerro Sosa (Montañas de Maquipucuna), which occupies most of the reserve, is extremely steep and dissected, although the main ridge slopes more gradually from the northwestern end at c. 1700 m elevation to the southeastern end in the vicinity of Cerro Montecristi (nearly 2800 m). Due to the proximity of Maquipucuna to Volcán Pichincha, it appears that the rock formations are mainly of Pleistocene age. Before the establishment of the reserve, there was some prospecting for bauxite, but fortunately commercial exploitation never materialized.

VEGETATION

The vegetation of Maquipucuna is prevailingly evergreen rain forest of considerable complexity; on ridges it remains relatively undisturbed, but below 1500 m it has mostly been drastically altered, converted to pastures or fields of crops such as banana, manioc, sugar cane, and "naranjilla" (Appendix A). The mosaic of areas with different times of cessation of disturbance results in secondary vegetation of considerable variability.

The northernmost and lowest points in the area covered by this inventory (north of the Maquipucuna reserve proper), at c. 1100 m, lie near the elevational boundary between lowland tropical forest and lower montane forest as delimited by Lauer (1986) and Webster (1995). The cloud forest in the reserve, above 1500 m, more or less corresponds in elevation to the *Cinchona* belt of Humboldt (1807) and the "selva pluvial mesotérmica" of Acosta-Solís (1966); it includes both "lower montane rain forest" and "upper montane rain forest" as defined by Grubb (1977) and Webster (1995). The Maquipucuna *lower montane cloud forest*, where well preserved, is a classic exemplar of the vegetation type, with a high diversity of tall straight trees up to 25–35 m high, many epiphytic ferns, bromeliads, and orchids, and characteristic cloud forest life forms such as tree ferns, palms, and festoons of bryophytes on the branches and logs (Appendix B). Above 2000 m, the forest becomes progressively smaller in stature and mossier; from 2400 to 2800 m it is designated here as *upper montane cloud forest*. Jørgensen and Ulloa (1994), who have critically reviewed the high montane forests and their environmental conditions, treat the zone above 2400 m as "montane forest". This upper montane forest contains many genera in common with the lower montane forest, but also includes temperate elements such as *Cornus, Coriaria, Escallonia, Hydrangea,* and *Viburnum* (Appendix C). In contrast to the generally intact upper montane forest, in the lower montane forest below 1500 m, where there has been extensive clearing for pastures and plantings of banana and naranjilla, disturbed areas are beginning to revert to scrub or secondary forest; the flora therefore includes a number of successional and ruderal species such as *Cecropia monostachya, Piper* spp., and *Trema micrantha* (Appendix A). South of the reserve headquarters, removal of livestock from an extensive field adjacent to the guava orchard has resulted within three or four years in a very dense stand of

Baccharis scrub (Zahawi & Augspurger, 1999). A characteristic riparian forest, at 1100–1300 m, disturbed to varying degrees but with large scattered trees of *Ceiba* and *Ficus*, often cloaked by the conspicuous liana *Sarcopera* [*Norantea*] *anomala*, occurs along the banks of the Alambi, Tulambi, and Umachaca rivers (Appendix D).

FLORISTICS OF
ECUADORIAN CLOUD FORESTS

The publication of this checklist of a cloud forest area in Ecuador is significant because of the scarcity of annotated floristic lists for cloud forest areas in the Andes as a whole and in Ecuador in particular. The only complete flora (with keys and illustrations) for *any* neotropical cloud forest area is that of Steyermark and Huber (1978; emended by Meier, 1998) for the Serra del Ávila in Venezuela, which does not have a very typical Andean flora. In Ecuador, published local floras or checklists of rain forest areas cover lowland (Dodson & Gentry, 1993) or mid-elevations below 1000 m (ENDESA reserve, Jørgensen & Ulloa, 1989). The recent manual of Andean trees and shrubs in Ecuador (Ulloa & Jørgensen, 1993) and the checklist of seed plants of the High Andes (Jørgensen & Ulloa, 1994) are very useful, but account only for taxa above 2400 m. Acosta-Solís (1966) has provided lists of woody plants at selected places in the "selva pluvial mesotérmica subandina", but neither in that work nor elsewhere is there a published checklist for any significant cloud forest area in Ecuador covering the 1000–2000 m elevational range. However, articles in the proceedings of the 1993 Neotropical montane forest symposium at the New York Botanical Garden (Churchill et al., 1995) have provided new data that greatly expedite comparisons of the Maquipucuna flora with comparable sites in Colombia and Peru.

Despite the paucity of floristic treatments of Pacific-slope Andean cloud forests, much botanical work has been done in Ecuador since the pioneering trip of Joseph Jussieu in La Condamine's expedition in the 1740's (Acosta-Solís, 1968; Renner, 1993; Jørgensen & León-Yánez, 1999). The "Nanegal Valley" (which referred to the area between the Alambi, Pichán, and Umachaca rivers) is sufficiently close to Quito that it was an early scene of botanizing. Juan Tafalla and his collaborators are recorded as visiting Nanegal about 1803, at the same time that Humboldt and Bonpland were making their historic researches in the highlands of Ecuador (Estrella, 1989). Probably the most extensive collections in the Nanegal region before the present time were made by Padre Luis Sodiro during the last quarter of the 19th century (Diels, 1937). Sodiro collected aroids in the Nanegal Valley at least as early as 1874 and as late as 1903, according to specimens cited by Engler (1905). The bromeliad enthusiast François André, although stationed at Nieblí (east of our boundary, near Pululahua), visited the

Nanegal Valley in 1876 in company with Sodiro (Padilla, 1983; Nicolson, 1983). Many of the species of ferns, aroids, and Piperaceae described by Sodiro (or based on his collections) are cited as from Nanegal or nearby localities. It is possible that some of Sodiro's species were actually collected within boundaries of the Maquipucuna reserve, but his indications of locality are too general to be sure; probably most of his specimens were taken along the Río Alambi and adjacent slopes (outside the reserve limits but still within our floristic area).

Within Ecuador, there are few published checklists of montane rain forest areas that could provide comparisons with the flora of Maquipucuna. As previously stated, the recent checklist of the high Andean flora by Jørgensen and Ulloa (1994) is quite useful for upper montane taxa (2400 m and above) at Maquipucuna, where there are 112 of their 137 "characteristic" genera. The study of Volcán Pasochoa by Jørgensen and Valencia (1988) cites only selected species from upper montane rain forest above 2800 m, beyond the elevational range of Maquipucuna. A number of common Maquipucuna plants of ethnobotanical interest are illustrated and discussed in a review of useful plants of Pichincha Province by Rios (1993). A small selection of characteristic species from Maquipucuna and adjacent areas is given by Sarmiento (1994). Of considerable relevance to the Maquipucuna area is the analysis of the useful plants of the Reserva Geobotánica del Pululahua by Cerón (1993), which enumerates 260 species (out of 650 collected by the author). The western boundary of the Pululahua Reserve lies only a few kilometers east of the eastern boundary of Maquipucuna, and not unexpectedly it shares a considerable number of species. However, most of the Pululahua Reserve is higher and drier than Maquipucuna, and it appears to have a less diverse flora. Perhaps the area most similar to Maquipucuna is the Reserva Florística Ecológica "Río Guajalito", some 40 km to the south; 345 species have been reported from this reserve of 400 hectares (Jaramillo & Grijalva, 1993). When a checklist of the Río Guajalito flora is published, it will permit significant comparisons with Maquipucuna.

As a consequence of the lack of published floras for montane forest areas in Ecuador, much of our knowledge rests on quadrat studies of 0.1–1 ha areas; Gentry (1995) gave some summary data on Maquipucuna based on his 0.1 ha quadrats. Vegetational and floristic structure in a site comparable to the upper part of the lower montane forest at Maquipucuna (2000 m) has been described in a 1 ha transect from the eastern slopes of the Andes near Baeza (Valencia, 1995). It is notable that among the 45 woody genera sampled in the Baeza plot, only four have not been recorded from Maquipucuna: *Bunchosia*, *Cedrela*, *Juglans*, and *Mollinedia*. Additional detailed quadrat studies are needed at Maquipucuna to supplement the unpublished surveys of Gentry and others.

Because of the elevational range within Maquipucuna of about 1600 m (1200–2800 m), it is of some interest to determine the extent that the various taxa occur at different elevations (Fig. 2). However, it should be kept in mind that elevational limits elsewhere in Ecuador may be different from those in Maquipucuna; detailed summaries of elevational limits are provided by Jørgensen and León-Yánez (1999). Not surprisingly, a number of families range throughout, from 1150 or 1200 to over 2700 m. However certain "lowland" taxa are much more restricted, for example, the woody legumes *Calliandra, Erythrina, Mimosa,* and *Mucuna,* which generally do not exceed 1750 m. The characteristic myrmecophyte tree *Cecropia* has been collected only to 1800 m and occurs rarely above that level. Other taxa restricted to lower elevations include the Acanthaceae, Agavaeae, Annonaceae, Apocynaceae, Chrysobalanaceae, Cyclanthaceae (except for *Sphaeradenia horrida*), Marantaceae, and Marattiaceae to 1800 m (or less); Lecythidaceae to 1900 m; *Equisetum, Ficus,* and *Hydrangea* to 2000 m; Asclepiadaceae to 2100 m; and Cyatheaceae (except for *Cyathea caracasana*) and Euphorbiaceae to 2300 m. In contrast, a number of "upland" taxa do not occur locally below 1400–1500 m: *Brunellia, Escallonia, Gunnera, Meliosma, Smilax,* and *Marcgravia. Arracacia, Cardamine, Chusquea, Clethra, Cornus,* and *Cortaderia* have not been encountered below 1950–2000 m. A number of "subalpine" genera have been recorded only above 2500 m: *Lepechinia, Mutisia,* and *Pernettya.*

A comparison with the trees listed in the recent tree flora (Cuamacás & Tipaz, 1995) of interandean forests in northern Ecuador (between Maquipucuna and the Colombian border) shows that in general elevational ranges are comparatively lower at Maquipucuna. For example, Aliso (*Alnus acuminata*) was observed at Maquipucuna as low as 1300 m (although more common at 2000 m and above), whereas Cuamacás and Tipaz list the species at 2300–3500 m for Ecuador. Similarly, *Hyeronima macrocarpa* has been found between 1800 and 2300 m at Maquipucuna, but 2800–3200 m in the sites studied by Cuamacás and Tipaz. Other examples include: *Escallonia paniculata,* 1675–2200 m at Maquipucuna vs. 2800–3500 m, *Ruagea tomentosa,* 1700–2200 m at Maquipucuna vs. 3000–3400 m, and *Turpinia occidentalis,* 1500–2300 m at Maquipucuna vs. 2800–3100 m. This difference in elevational ranges, also noted for the palms of Maquipucuna by Svenning and Balslev (1998), may be correlated with the higher annual rainfall at Maquipucuna (possibly due to the Massenerhebung effect according to Svenning & Balslev). However, in contrast to this general trend, some taxa reported from lowlands by Jørgensen and León-Yánez (1999) in Ecuador have been found at much higher elevations in Maquipucuna, e.g., *Apodanthera biflora* (Cucurbitaceae) at 1850–2250 m, and *Eschweilera integrifolia* (Lecythidaceae) at 1550–1700 m. Much more intensive sampling is still needed at Maquipucuna in order to establish more definitely the exact upper and lower elevational ranges of the various taxa.

At present, the flora of Maquipucuna appears to be moderately well sampled within the core area of the reserve; probably at least half of the species growing within the floristic area have now been collected (although a sizeable number of collections remain unidentified). The pteridophytes, which have received the most attention, are the most thoroughly studied group; perhaps up to three quarters of their species in the area have now been recorded. On the other hand, epiphytes, including Araceae, Bromeliaceae, Orchidaceae, and *Peperomia* remain inadequately sampled. For example, Dodson and Escobar (1996) have cited 306 species of Orchidaceae from a 10 square kilometer area in the superwet area of Lita; we suspect that the 200 species recorded at Maquipucuna, from 15 square kilometers, may represent scarcely more than half of the orchid flora. A number of families known from mid-elevation Andean montane forests have not yet been recorded from Maquipucuna; the most outstanding examples include: Anacardiaceae, Aristolochiaceae, Berberidaceae, Burmanniaceae, Buxaceae, Cactaceae, Eleaocarpaceae, Hydrophyllaceae, Juglandaceae, Lentibulariaceae, Myricaceae, Oleaceae, Podocarpaceae, Rhamnaceae, and Sterculiaceae. Further collecting may verify the presence at Maquipucuna of some of these families and many genera yet unrecorded; for example, *Salacia* (Hippocrateaceae) and *Clavija* (Theophrastaceae) were not collected until 1997.

ANALYSIS OF THE
VASCULAR FLORA OF MAQUIPUCUNA

The recorded flora of the Maquipucuna area, as geographically delineated in this checklist, currently includes 621 genera and 1,640 species of vascular plants, of which only 44 are exotic. The 586 native genera represent 27.4% of the native Ecuadorian genera and the 1,596 native Ecuadorian species represent 10.4% of the native species of Ecuador reported by Jørgensen and León-Yánez (1999). Tables 1–6 summarize some of the statistics of the flora, which falls entirely within the mid-elevation range of the Ecuadorian Andes as defined by Balslev (1988); the lowermost points in the flora area, at c. 1100 m, and in the Maquipucuna reserve (c. 1200 m), are well above the upper edge of the lowland/upland transition belt (900 m) recognized by Balslev and Renner (1989). It is notable that five of the 10 largest families (Table 2), and six of the 10 largest genera (Table 3), are pteridophytes or monocots. The family representation of the Maquipucuna flora on the whole appears to be typical for the lower montane forest in Ecuador; of the 10 largest families (Table 2), only one, Solanaceae, is lacking from the summary by Jørgensen and León-Yánez (1999) of the most species-rich families at 1000–1500 m. For woody species the analysis of Gentry (1995; fig 5) showing dominance at mid-elevations (1500–1700 m) of Lauraceae, Melastomataceae, Rubiaceae, Moraceae, and Fabaceae, is in distinct contrast with Maquipucuna (Table 4), in which the largest woody families (by rank) are Rubiaceae, Asteraceae, Solanaceae, Melastomataceae, and Piperaceae. The relatively low ranking of Moraceae, Clusiaceae, and Euphorbiaceae at Maquipucuna suggests that in Maquipucuna "thermophilic" elements become attenuated more rapidly with increasing elevation than is typical for Andean cloud forests in general. This surprising discrepancy may arise partly from the Chocó elements in the Maquipucuna flora. The largest family in number of species, with about 12.8% of the flora, is undoubtedly the Orchidaceae, for which 204 species are reported currently (but at least another 50 await identification). The pteridophytes, with 227 native species (14.2% of the vascular flora), are a more prominent floristic component at Maquipucuna than at most other cloud forest sites for which data are available (Tables 7, 8, 9, 10, 13); the percentage of pteridophytes is notably larger than that for Ecuador as a whole (8.5%; Jørgensen & León-Yánez, 1999).

10

Table 1. Numbers of vascular plant taxa at Maquipucuna.

	Families		Genera		Species	
	Total	Native	Total	Native	Total	Native
Pteridophytes	20	20	55	54	228	227
Dicots	111	107	409	385	961	931
Monocots	24	24	157	147	451	438
Total	155	151	621	586	1640	1596

The most species-rich taxa clearly reflect the importance of epiphytes; about 441 species, or approximately 27.7% of those recorded from Maquipucuna, are epiphytic (Appendix E); this is only slightly more than the average (25.8%) for Ecuador as a whole (Jørgensen & León-Yánez, 1999). It should be kept in mind that although some taxa (Grammitidaceae, Vittariaceae, pleurothallid orchids) are mostly obligate epiphytes, there is a great deal of facultative epiphytism, so assignment of particular taxa is sometimes arbitrary. According to the definitions of Williams-Linera and Lawton (1995), 361 species are *holoepiphytes*, without any ground connection during their life cycle. The remaining 80 are *hemiepiphytes*, either germinating on branches in the canopy and later establishing roots in the ground (primary hemiepiphytes such as "stranglers" in *Ficus*), or germinating in the ground and climbing into trees (secondary hemiepiphytes such as Marcgraviaceae, Cyclanthaceae, and some aroids). Although not listed by Williams-Linera and Lawton, genera such as *Begonia* may prove to have hemiepiphytic species; critical observations on epiphyte morphology and ecology are clearly needed. It is notable that five of the ten largest families at Maquipucuna have significant numbers of epiphytic species (Table 2). Of the 10 largest genera (Table 3), only two (*Solanum* and *Psychotria*) lack epiphytic species. Although the Polypodiaceae have the highest percentage of epiphytism (87%) of the larger families, the most epiphyte-rich elements of the Maquipucuna flora are the orchids and Araceae, both families with over 50 epiphytic species. Our most species-rich genus, *Peperomia*, has at least 17 epiphytic species out of 41. Despite some prevailingly epiphytic families such as Polypodiaceae and Grammitidaceae, the pteridophytes overall have only 47% epiphytism (106 out of 227 species epiphytic).

Table 2. Diversity of the 16 most species-rich vascular plant families at Maquipucuna. Only native taxa are tabulated.

Family	Number of genera	Number of species	Epiphytic species
Orchidaceae	66	204	150
Asteraceae	52	84	0
Piperaceae	2	79	19
Araceae	8	69	51
Rubiaceae	22	66	1
Solanaceae	17	59	4
Dryopteridaceae	12	51	23
Melastomataceae	12	40	0
Poaceae	22	35	0
Gesneriaceae	9	35	18
Bromeliaceae	6	35	27
Cyperaceae	9	31	0
Polypodiaceae	5	30	26
Fabaceae	14	29	0
Ericaceae	9	26	13
Lauraceae	8	25	0

The prevalence of aroids in the Maquipucuna flora is especially notable. *Anthurium*, with 40 species, is by far the largest monocot genus in our flora. For the recording of this diversity, we are particularly indebted to the Jesuit botanist Luis Sodiro (1903), who described 30 of the Maquipucuna species (13 reduced to synomy in CVPE) in a series of publications (Nicolson, 1983). An indication of the high degree of endemism in *Anthurium* is that 17 of these Sodiro species were described (types or syntypes) from the Nanegal Valley. At least 10 of these Sodiro species appear not to have been recollected; because of the extensive clearing in the Nanegal (Alambi) Valley, some of them are possibly now extinct, although this needs to be verified by intensive field work along the western edge of the Maquipucuna area.

Table 3. Genera of vascular plants at Maquipucuna with the largest number of native species and the number of epiphytic species within these genera.

Genus	Native species	Epiphytic native species
Peperomia	41	17
Anthurium	40	35
Piper	38	2
Epidendrum	31	20
Solanum	23	0
Asplenium	19	13
Elaphoglossum	18	16
Psychotria	17	0
Philodendron	16	15
Guzmania	16	14
Diplazium	16	3
Begonia	15	6
Thelypteris	15	1
Miconia	15	0
Maxillaria	14	11
Polypodium	14	10
Pilea	14	3
Pleurothallis	13	13
Ficus	13	0
Passiflora	13	0
Palicourea	13	0
Huperzia	12	6
Elleanthus	12	6
Hymenophyllum	11	11
Columnea	11	11

Table 4. Diversity of woody taxa in the most prominent families at Maquipucuna.

Family	Genera with woody species	Native woody species
Rubiaceae	16	59
Asteraceae	23	53
Solanaceae	13	46
Piperaceae	1	37
Melastomataceae	10	36
Lauraceae	8	27
Ericaceae	9	26
Gesneriaceae	6	25
Euphorbiaceae	9	24
Moraceae	6	19
Fabaceae	10	18
Clusiaceae	7	15
Myrtaceae	5	14
Myrsinaceae	5	11
Araliaceae	3	11

The occurrence of so many aroids described from the Nanegal region raises the question of whether the species mentioned above are endemic to our area (Sodiro, 1903). So far, the most distinctive endemic species appear to be *Ardisia websteri* (Pipoly, 1996) and *Clibadium websteri* (Robinson, 1997) which have not yet been recorded anywhere else. A number of epiphytic orchids have not been collected elsewhere, except sometimes in the adjacent Pululahua or Gualea areas.

It should be noted that the Maquipucuna flora is also rich in arboreal taxa; the total of 125 genera and 310 species of Maquipucuna trees is greater than any of the cloud forest stations summarized by Vázquez-García (1995). The species representation by family at Maquipucuna for the 1000–2000 m zone is similar to that reported for Ecuador as a whole by Jørgensen and León-Yánez (1999), with the striking exception that at Maquipucuna the Melastomataceae and Mimosaceae are relatively less well represented. It is surprising that the palms are represented by only nine genera with 11 species. Both canopy trees and epiphytes are probably underrepresented in the statistics of the flora, largely because of difficulties in collecting material.

Table 5. Genera of vascular plants at Maquipucuna with the largest number of woody species.

Genera	Number of woody species
Piper	37
Solanum	17
Miconia	15
Ficus	14
Palicourea	14
Psychotria	14
Ocotea	9
Cyathea	8
Tournefortia	8
Macleania	7
Alloplectus	7
Nectranda	7
Clusia	7
Rubus	7
Baccharis	6
Casearia	6
Clibadium	6

Another striking feature of the Maquipucuna flora is the small number of introduced species (Table 6): only 44 have been recorded (less than 3% of the flora, omitting strictly cultivated species and eight "pantropical" species); three commonly cultivated species (*Bixa orellana*, *Coffea arabica*, and *Psidium guajava*) are only dubiously naturalized. There are serious problems in determining what the "native" status means; among some 65 herbaceous weedy species recorded from our area, there are a considerable number (especially in Asteraceae and Fabaceae) that are native in the broad sense of being neotropical, but may be recently introduced into the Maquipucuna area. The grasses, with 11 introduced species, have the most exotics, but it is interesting that the family Fabaceae, which is relatively poorly represented in the flora, is second with eight species. Overall, the lack of invasive species in intact rain forest is notable. In mature or lightly disturbed secondary forest, only two exotic species are commonly seen: *Macrothelypteris torresiana* and *Oplismenus burmannii*; in primary forest, no exotics are evident at all.

Table 6. Exotic taxa in the Maquipucuna flora, arranged by family.

Macrothelypteris torresiana
Amaranthus hybridus
Amaranthus viridus
Gnaphalium luteo-album
Hypochaeris radicata
Taraxacum officinale
Impatiens wallerana
Bixa orellana
Cynoglossum amabile
Heliotropium indicum
Hippobroma longiflora
Chenopodium ambrosioides
Bryophyllum pinnatum
Lagenaria siceraria
Manihot esculenta
Ricinus communis
Medicago polymorpha
Melilotus indica
Pachyrhizus tuberosus
Pueraria phaseoloides
Spartium junceum
Trifolium repens

Vigna unguiculata
Centaurium erythraea
Persea americana
Psidium guajava
Syzygium jambos
Oenothera pubescens
Plantago major
Rumex obtusifolius
Nicandra physalodes
Anthoxanthum odoratum
Briza minor
Coix lacryma-jobi
Cymbopogon citratus
Digitaria abyssinica
Eleusine indica
Holcus lanatus
Oplismenus burmannii
Pennisetum purpureum
Poa annua
Setaria sphacelata
Vulpia myuros
Hedychium coronarium

Note: Status of several taxa is unclear because it is not certain to what extent cultivated species are truly naturalized. Some weedy species (especially Asteraceae and Poaceae) are arbitrarily treated as native even though they occur in both the Old World and neotropics.

RELATIONSHIPS WITH OTHER FLORAS

The data presented by Balslev (1988) suggest that in Ecuador the western slopes of the Andes may have both the largest number of species and the highest percentage of endemism in the vascular flora. Floristically, Maquipucuna falls within the Northwestern Region of the high Andean floras as delineated by Jørgensen and Ulloa (1994). However, the great diversity and distinctive composition of the Maquipucuna flora indicates that it may be regarded as a mid-elevation Andean extension of the Colombia Chocó forest flora (Gentry, 1986). In fact, the checklist of the flora of Cerro del Torrá in the Chocó (Silverstone-Sopkin & Ramos-Pérez, 1995) shows a greater similarity to the flora of Maquipucuna than any florula published so far. Table 7 provides a comparison of the floras of Maquipucuna, Cerro del Torrá, and Chocó province (Forero & Gentry, 1989). Except for the inclusion of the Fabaceae in Chocó (doubtless reflecting lowland species), the list of the most species-rich families is strikingly similar for the three floras. Maquipucuna and Cerro del Torrá appear somewhat anomalous in the high representation of Asteraceae (5.3% and 3.6% respectively), while Maquipucuna has a relatively low complement of Melastomataceae (2.5%), and the Araceae do not appear in the list at Cerro del Torrá. Nevertheless, the similarities are more striking than the differences, particularly when compared with drier or lower elevational sites from elsewhere. Maquipucuna appears marginally northeastern and upslope from the Tumbesian region as defined by Kessler (1995) for areas in southern Ecuador and northern Peru; the floristic relationships between the Chocó and Tumbesian regions need critical study, but it is clear the Maquipucuna region is much more humid and floristically richer than corresponding Tumbesian regions.

Two recent detailed studies of neotropical cloud forests in Monteverde, Costa Rica (Haber, 2000), and in Parque Ávila, Venezuela (Meier, 1998) permit detailed comparisons with the Maquipucuna flora (Table 8). The cloud forest of Monteverde, at 1200–1700 m, is closely comparable to the lower montane cloud forest at Maquipucuna: the 12 largest families are identical except that the Gesneriaceae in Maquipucuna are replaced by the Fabaceae in Monteverde. The extraordinarily high number of orchid species at Monteverde, although including some lowland species, indicates both the richness of that site and, quite likely, a

Table 7. Comparison of the Maquipucuna flora with the floras of the Cerro del Torrá (Silverstone-Sopkin & Ramos-Pérez, 1995) and the Provincia de Chocó (Forero & Gentry, 1989). Numbers indicate species totals for the largest families (ferns and legumes treated as single units) and percentages of the total flora. Maquipucuna totals and percentages include only native species.

Maquipucuna			Cerro del Torrá			Prov. Chocó		
Family	Number of species	Percentage of total flora	Family	Number of species	Percentage of total flora	Family	Number of species	Percentage of total flora
Orchid.	204	12.8	Ferns	75	16.0	Ferns	428	11.1
Ferns	202	12.7	Orchid.	46	9.8	Orchid.	335	9.7
Aster.	84	5.3	Melast.	40	8.5	Rubiac.	268	6.9
Piper.	79	4.9	Rubiac.	35	7.5	Fabac.	183	4.7
Arac.	69	4.3	Ericac.	30	6.4	Melast.	181	4.7
Rubiac.	66	4.1	Gesner.	23	4.9	Piper.	140	3.6
Solan.	59	3.7	Aster.	17	3.6	Aster.	106	2.7
Melast.	40	2.5	Bromel.	15	3.2	Gesner.	95	2.5
Bromel.	35	2.2	Arac.	14	3.0	Arac.	91	2.4
Gesner.	35	2.2	Piper.	14	3.0	Solan.	72	1.9
Poac.	35	2.2	Solan.	14	3.0	Euphorb.	72	1.9

Table 8. Comparison of the Maquipucuna flora with the floras of Monteverde (Haber, 2000) and Parque Ávila (Meier, 1998). Numbers indicate species totals for the largest 12 families (ferns and legumes treated as single units), and percentages of the total flora. Maquipucuna totals and percentages include only native species; Monteverde totals are based on records from 1200–1700 m (except for Orchidaceae, which also includes lowland species).

	Maquipucuna			Monteverde			Ávila		
Family	Family	Number of species	Percentage of total flora	Family	Number of species	Percentage of total flora	Family	Number of species	Percentage of total flora
	Orchid.	204	12.8	Orchid.	[393]	13.0	Pterid.	237	10.5
	Ferns	202	12.7	Ferns	208	12.2	Orchid.	182	8.0
	Aster.	84	5.3	Aster.	98	5.7	Poac.	167	7.4
	Piper.	79	4.9	Rubiac.	90	5.3	Fabac.	158	7.0
	Arac.	69	4.3	Solan.	65	3.8	Aster.	127	5.6
	Rubiac.	66	4.1	Piper.	49	2.9	Rubiac.	86	3.8
	Solan.	59	3.7	Melast.	48	2.8	Solan.	63	2.8
	Melast.	40	2.5	Poaceae	47	2.8	Euphorb.	57	2.5
	Bromel.	35	2.2	Bromel.	45	2.6	Melast.	51	2.3
	Gesner.	35	2.2	Laurac.	41	2.4	Bromel.	48	2.1
	Poac.	35	2.2	Fabac.	38	2.2	Piper.	43	1.9
	Cyper.	31	1.9	Arac.	36	2.1	Arac.	35	1.5
Species Total	1596			1708			2264		

more thorough inventory of the flora by Haber and his associates. The exceptional diversity of Monteverde is further evident when it is recalled that the totals for aquipucuna include upper montane species as well as lower montane. The only families distinctly better represented at Maquipucuna are the Araceae (scarce in upper montane forest), Gesneriaceae, and Piperaceae; Monteverde appears definitely richer in Rubiaceae.

The flora of Ávila is very different in a number of respects, since it includes lowland species to near sea-level, and a larger area than either Maquipucuna or Monteverde. Furthermore, since it is close to the Venezuelan capital, Caracas, the vegetation at Ávila is much more disturbed; this is particularly reflected in the extraordinarily high percentage of grass species. As pointed out by Meier (1998), the very high percentage of legumes at Ávila reflects their relative dominance at lower elevations. Nevertheless, Ávila to some extent shares with Maquipucuna and Monteverde high percentages of Asteraceae, Rubiaceae, Solanaceae, and Melastomataceae, as well as the epiphyte-rich familes Araceae, Bromeliaceae, and Piperaceae. Although a detailed comparison at the genus and species level of the taxa at the three sites has not yet been undertaken, it appears that the inventories have reached a level of sampling adequate to support critical statistical analyses. A casual inspection of the three checklists suggests, not unexpectedly, that there is a very high percentage of families and genera in common, but (except in pteridophytes) a very low percentage of species in common.

The richness of the Maquipucuna flora is strikingly illustrated by the comparison in Table 9 of species numbers compared to the entire eastern mid-elevation Andean slopes in Peru (Young, 1991). There is a notable relative poverty of the Peruvian cloud-forest belt in ferns, gesneriads, and aroids. An even more significant comparison is that of Maquipucuna with the drier cloud forests of northwestern Peru (Dillon et al., 1995); the five sites combined have a flora (1100 species) comparable in size to that of Maquipucuna. The relatively high numbers of Asteraceae, Solanaceae, and Poaceae, contrasted with the diminution of Araceae, Ericaceae, and Gesneriaceae in northwestern Peru, in all probability reflects the lower rainfall in that area.

It is now possible to compare the flora of Maquipucuna with that of a distant montane rain forest area in Brazil: Macaé de Cima, in the Serra do Mar (Lima & Guedes-Bruni, 1994; Table 10). Much of the Reserva Ecológica de Macaé de Cima lies at a lower elevation (880–1720 m). The overall size of the recorded flora (973 species) is smaller than that of Maquipucuna, and it is apparent from Table 10 that the floristic composition is very different. Aside from the ferns, the two floras have only one family—Rubiaceae—in common among the five largest. The flora of Macaé de Cima has a remarkably greater proportional representation of Melastomataceae, Fabaceae, Myrtaceae, and Lauraceae; it is much lower in Asteraceae, Orchidaceae, Piperaceae, Araceae, Gesneriaceae, and Ericaceae. Because of the scarcity of published florulas from other sites in the Andes and the

Table 9. Comparison of the most species-rich families in the Maquipucuna flora with their representatives in the summary flora for eastern slopes of the Peruvian Andes between 1500 and 2500 m (Young, 1991) and for the northwestern Peruvian cloud forests (Dillon et al., 1995). Maquipucuna totals include only native species.

Family	Maquipucuna Number of species	Eastern Peru Number of species	Northwestern Peru Number of species
Orchidaceae	204	202	59
Asteraceae	84	25	78
Piperaceae	79	44	39
Araceae	69	40	12
Rubiaceae	66	78	40
Solanaceae	59	43	55
Dryopteridaceae	51	36	22
Melastomataceae	40	66	39
Bromeliaceae	35	33	35
Gesneriaceae	35	35	11
Poaceae	35	35	48
Polypodiaceae	30	10	27

Serra do Mar, it is not clear to what extent these differences reflect environmental or historical factors. The high proportion of Fabaceae at Macaé de Cima may possibly be explained by its lower elevation and somewhat drier climate. However, the great abundance of Myrtaceae and Lauraceae in the Brazilian reserve may provide an indication of the different floristic histories of the Andes and the Serra do Mar.

The distinctive floristic characteristics of Maquipucuna as a montane area are indicated in comparisons with the flora of Río Palenque (Dodson & Gentry, 1978) in the Ecuadorian Pacific lowlands (Tables 11–13). Notable differences (Table 11) include the much greater representation of ferns, Asteraceae, Ericaceae, and Rubiaceae at Maquipucuna, whereas Río Palenque (Table 13) has a more impressive total of Fabaceae (*s. lat.*), Apocynaceae, and Moraceae, reflecting the general floristic pattern for lowland as opposed to upland rain forest areas (Gentry, 1992). Some "thermophilic" families at Río Palenque, which mainly occur below 1000 m, have not been found at Maquipucuna, e.g., Aristolochiaceae, Cactaceae, Hernandiaceae, Podostemonaceae, and Sterculiaceae. Overall, however, the

Table 10. Comparison of the most species-rich families in the flora of Maquipucuna with those of Macaé de Cima (Rio de Janeiro, Brazil).

Maquipucuna		Macaé de Cima	
Family	Number of species	Family	Number of species
Orchidaceae	204	Ferns	88
Ferns	202	Melastomataceae	60
Asteraceae	84	Fabaceae	50
Piperaceae	79	Myrtaceae	47
Araceae	69	Lauraceae	45
Rubiaceae	66	Rubiaceae	45
Solanaceae	59	Asteraceae	40
Melastomataceae	40	Bromeliaceae	37
Bromeliaceae	35	Orchidaceae	35
Gesneriaceae	35	Solanaceae	30
Poaceae	35	Bignoniaceae	20

Table 11. Comparison of species numbers of the largest families in the Maquipucuna flora with the species numbers of these families in lowland rainforest at Río Palenque (Dodson & Gentry, 1978).

Family	Maquipucuna Number of species	Río Palenque Number of species
Orchidaceae	204	127
Asteraceae	84	46
Piperaceae	79	43
Araceae	69	40
Rubiaceae	66	27
Solanaceae	59	37
Dryopteridaceae	51	20
Melastomataceae	40	20
Bromeliaceae	35	18
Gesneriaceae	35	32
Poaceae	35	15
Cyperaceae	31	23

Table 12. Comparison of numbers of woody species in families at Maquipucuna with Río Palenque (Dodson & Gentry, 1978).

Family	Maquipucuna Number of woody species	Río Palenque Number of woody species
Rubiaceae	59	23
Asteraceae	53	15
Solanaceae	46	28
Piperaceae	37	27
Melastomataceae	36	16
Lauraceae	27	12
Ericaceae	26	5
Euphorbiaceae	24	19
Moraceae	19	39
Fabaceae	18	29

Table 13. Comparison to indicate families better represented by native species at Río Palenque (Dodson & Gentry, 1978) than at Maquipucuna.

Family	Maquipucuna Number of species	Río Palenque Number of species
Fabaceae	29	42
Moraceae	20	37
Arecaceae	11	14
Bignoniaceae	7	10
Acanthaceae	5	17
Menispermaceae	5	8
Apocynaceae	4	9
Malpighiaceae	1	6
Myristicaceae	1	5
Sapotaceae	1	6

Table 14. Relative rank of families in species number for Maquipucuna compared with lowland rain forest floras from La Selva, Costa Rica (Hammel, 1990) and from Cocha Cashu, Peru (Gentry, 1990). Ferns are here treated as a "family" only for comparison. Maquipucuna totals include only native species.

Maquipucuna		La Selva		Cocha Cashu	
Family	Number of species	Family	Number of species	Family	Number of species
Orchidaceae	204	Ferns	169	Fabaceae	102
Ferns	202	Orchidaceae	114	Moraceae	66
Asteraceae	84	Araceae	99	Rubiaceae	65
Piperaceae	79	Rubiaceae	99	Ferns	65
Araceae	69	Fabaceae	79	Orchidaceae	45
Rubiaceae	66	Piperaceae	79	Acanthaceae	44
Solanaceae	59	Melastomataceae	71	Sapindaceae	40
Melastomataceae	40	Poaceae	63	Bignoniaceae	38
Bromeliaceae	35	Asteraceae	48	Araceae	38
Gesneriaceae	35	Euphorbiaceae	42	Solanaceae	38
Poaceae	35	Cyperaceae	32	Myrtaceae	37

resemblances are rather striking, and the Río Palenque appears more similar to Maquipucuna than to the other neotropical lowland stations discussed by Gentry (1990).

In comparison (Table 14) with the lowland floras of La Selva, Costa Rica (Hammel, 1990) and Cocha Cashu, Peru (Gentry, 1990), Maquipucuna has a strikingly high proportion of Asteraceae, Ericaceae, and Gesneriaceae, and a relatively low representation of Fabaceae, Moraceae, Acanthaceae, Bignoniaceae, and Sapindaceae. However, what is also notable is the unexpectedly great floristic similarity between Maquipucuna and La Selva (especially in the high percentages of Araceae and Piperaceae) as compared to Cocha Cashu. Hammel (1990) has noted that the lowland La Selva flora has an anomalously "montane" component rich in epiphytic taxa, which accounts for the resemblance with Maquipucuna; one could with little exaggeration regard La Selva and Maquipucuna as the polar extensions of the Chocó flora. When the Maquipucuna flora becomes better known, comparisons at the level of genus and species will permit more meaningful comparisons with neotropical upland and lowland rain forest areas.

As pointed out by Ulloa and Jørgensen (1993), the upper montane woody Andean flora of Ecuador is composed of a number of floristic elements. Perhaps because of its mostly lower elevation, the Maquipucuna flora has a rather sparse representation of their Tropical Andine element: only eight of 45 genera. In contrast, the Neotropical element is much better represented, with 55 of 113 genera in their enumeration, and by 17 of the 23 "thermophilic" genera that they note as penetrating upper montane forest. It is not surprising that the cloud forests of Maquipucuna have 13 of the 17 genera enumerated for the American-Asiatic element, but it is somewhat unexpected that the Austral-Antarctic element is so poorly represented by only five genera out of 16.

Overall, the Maquipucuna flora appears typical of both North and South American cloud forests in its elevated percentage of families with a high proportion of epiphytes: Araceae, Orchidaceae, Piperaceae, and pteridophytes. It is anomalous in comparison to sites in other countries in its unusually high representation of Asteraceae, Piperaceae, and Solanaceae, but not when compared to mid-elevation regions in Ecuador (Jørgensen & León-Yánez, 1999). This suggests a historical regional effect, perhaps related with the high percentage of endemism in Ecuador, but no causal explanations for this have been given.

CONSERVATIONAL STATUS OF THE MAQUIPUCUNA CLOUD FORESTS

The plant diversity documented by this checklist indicates that Maquipucuna is one of the botanical "crown jewels" of the Andes. Along the trails up Cerro Sosa and other ridges, the cloud forests surprise and delight with their cornucopian anarchy of tree ferns, gesneriads, orchids, and aroids, many perched inaccessibly in the crowns of a great variety of tall straight trees. Hikes in Maquipucuna become botanical detective investigations; many taxa (such as the local magnolia, *Talauma*) were first detected by flowers scattered along the trail. A large number of trees remain unidentified because flowers have never been found, and because of the difficulties of the precipitous terrain many trees have not yet been collected. A sighting of *Morpho* butterflies and quetzals along the trail is a reminder of the impressive number of insects and birds that also find sanctuary here.

The reviews of the history of the "Yumbo" peoples in northwestern Pichincha by Reyes (1993) and Saloman (1997) indicate that towns such as Nanegal may date back to pre-Incan times, and that other settlements along the Río Alambi such as Alambi and Cachillacta existed in the 16th century. The "Inca trail" that crosses the Maquipucuna reserve may well be a remnant of a pre-Incan trail connecting the Yumbo settlements with Quito. Clearing and deforestation along the Río Alambi almost surely antedates the 18th century, so the secondary forests on slopes below 1500 m in Maquipucuna may have been disturbed for centuries.

Although there is so far no evidence of massive extinction at Maquipucuna such as that documented for the lower-elevation Centinela forests by Dodson and Gentry (1993), it is notable that a number of the species of Araceae and Piperaceae described before 1910 by Sodiro have not been recollected. Consequently, there is reason for concern for the survival of some rarer Maquipucuna species if disturbance continues unabated.

Maquipucuna is one of the 61 neotropical protected sites included in the Parks in Peril program of the Nature Conservancy (Mansour, 1995). The present conservation status of the Maquipucuna region has been reviewed by Sarmiento (1994, 1995), who notes that the disturbance of the local ecosystems has been increasing due to proximity to Quito and improved access along the road from Calacalí to Nanegalito. An ecological and socioeconomic assessment of the Maquipucuna Reserve and its buffer zone—Evaluación Socio-ambiental Rápida (ESAR)—funded by IUCN is nearing publication (Rebeca Justicia, pers. com.).

The information gathered through ESAR will be used in preparing a management plan for Maquipucuna and the surrounding Guayllabamba Upper Watershed Protected Forest (GUWPF). As noted by Justicia and Sarmiento, environmental education programs are needed to obtain the necessary support from local communities.

Preservation of the cloud forests of Maquipucuna would be greatly facilitated by the implementation of the plan for the Chocó-Andean Biological Corridor (CABCOR) articulated by Justicia (1995). This would permit survival of more or less continuous forested areas from the Maquipucuna and Mindo reserves on the Andean slopes to the forests in the Cotacachi-Cayapas Ecological Reserve, with a planned extension to the lowland forests in Esmeraldas Province. We hope that publication of this checklist of the Maquipucuna flora will provide a convincing botanical argument in favor of creating the Chocó-Andean biological corridor and thus creating one of the world's most impressive reserves of plant diversity.

CIRCUMSCRIPTION OF THE REGION
COVERED BY THE INVENTORY

The floristic area of Maquipucuna, as defined for this inventory, lies within Cantón Quito, Provincia Pichincha; it includes not only the Bosque Protector Maquipucuna but also adjacent areas, including the Alambi Valley down to the 1100 m contour, Cerro Campana, Cerro Cachillacta, Cerro Negro, and the slopes above the highway immediately south of the reserve, extending to the equator (00°00′) just south of Tandayapa; for convenience, all of the land east of the road between Tandayapa, Nanegalito, and Nanegal has been included (Fig. 1). The latitudinal limits are 00°00′ to 00°10′N, and the longitudinal limits are approximately between 78°35′ and 78°41′ W; the area is entirely included within the Calacalí topographic sheet (1:50,000, Inst. Geogr. Militar). The total area included is approximately 22,000 hectares, with an elevational range of 1100–2800 m. However, only the approximately 5,000 hectares in the Maquipucuna reserve and immediately adjacent areas (Cerro Negro and the Loma Pahuamba area along the highway to Calacalí) can be regarded as adequately botanized, so that this checklist reflects primarily an inventory of the Maquipucuna core area. There are some problems with locality names, as successive topographic maps at different scales have been inconsistent. We have arbitrarily designated localities in the map so as to avoid duplication of locality names such as Cerro Campana and Cerro Sosa.

ENUMERATION OF TAXA

This checklist is primarily based on the botanical collections made by Grady L. Webster and collaborators on the UREP expeditions, which have been deposited at the University of California, Davis (DAV), the Herbario Nacional (QCNE), and the Pontificia Universidad Católica (QCA); additional duplicates have been distributed to other herbaria (including AAU, MO, NY, TEX, UC, US). Additional collections made by other botanists who contributed sizeable sets of specimens are cited with the collection number preceded by the appropriate letter: C (Carlos Cerón), D (Piero Delprete), F (Efrain Freire et al.), G (Alwyn Gentry et al.), H (Fred Hrusa), K (Dean Kelch), L (Bernt Lojtnant & Ulf Molau), M (Robbin Moran), N (David Neill), P (Walter Palacios), Q (Carlos Quelal), T (Galo Tipaz), V (Henk van der Werff), W (Kenneth A. Wilson), and Z (Zak Zahawi); other collectors are listed by surname. These specimens are mainly deposited at DAV, MO, and QCNE. An important collection of trees, shrubs, and epiphytes, made in the Loma Pahuamba area by Efrain Freire and collaborators, is housed at the Herbario Nacional (QCNE) in Quito. A number of other records are cited from the literature (especially treatments in the *Flora of Ecuador*) or from internet resources (mostly the W3TROPICOS database at the Missouri Botanical Garden). The total number of UREP collections is 3,625; together with the records of other collectors, the grand total of all collections made in the Maquipucuna area is over 5,200.

Nomenclature and delimitation of taxa mostly follow the *Catalogue of the Vascular Plants of Ecuador* (Jørgensen & León-Yánez, 1999), which is referred to as CVPE in the text; references have also been checked against the *Flora of Ecuador* and (for flowering plants) the checklist for the High Andes of Ecuador (Jørgensen & Ulloa, 1994). Synonyms are cited mainly in instances where the name adopted is not the same as in Jørgensen and León-Yánez (1999). Names of native species are in italics, introduced species are in underlined roman. Common names, indicated by quotation marks, are mostly Spanish; they have been compiled from herbarium specimen labels, personal contacts, and from floristic works on western Ecuador (Acosta-Solís, 1962; Cerón, 1993; Cerón & Ávila, 1994; Dodson & Gentry, 1978; Gentry, 1993; and Rios, 1993). About 50 species listed in square brackets are not numbered or counted in statistical summaries; these are mostly species collected slightly outside the boundaries of the floristic area (within 1' or 2').

Unless otherwise indicated, localities are mainly within the core area of Maquipucuna (from the headquarters at Hacienda El Carmen along the main ridge of Cerro Sosa [Montañas de Maquipucuna]) to the peak and slopes of Cerro Montecristi. For each species, the habitat is given, often in abbreviated form (primary forest: prim. for.; secondary forest: sec. for.; upper montane forest: upper mont. for.; riparian forest: rip. for.). Range limits in meters represent records from herbarium specimens and are generally rounded off to the nearest 25 or 50 m. Colors of inflorescences (inflors.), flowers (fls.), and fruits (frs.) are given based on specimens and personal observations. Plants not indicated as epiphytes (ep.) are assumed to be terrestrial (terr.). The total number of genera found in each family is indicated in parentheses after the family name. Species not reported from Provincia Pichincha in Jørgensen and León-Yánez (1999) are indicated by having their species number in bold type.

Determinations of specimens collected by G. L. Webster and other UREP personnel that were not identified by specialists (marked by one or more asterisks) were made by G. L. Webster and UREP collaborators (especially Dean Kelch, Montserrat Rios, Brian Smith, and Kenneth A. Wilson). Many collections remain unidentified to genus or to family; these have not been cited.

Upon publication, this checklist will be made available on the Internet via the University of California, Davis Herbarium web page, http://herbarium.ucdavis.edu, and will be periodically updated as exploration of the Maquipucuna area progresses.

PTERIDOPHYTA (reviewed by A. R. Smith)

EQUISETACEAE (1)

1. *Equisetum bogotense* Kunth
   Common in wet places, especially along streams, 1250–1950 m: 27226, 27229, 27230, 27509, 27590, 29088; W 2599. "Cola de caballo"

2. *Equisetum giganteum* L.
   Wet slopes, 1200–2000 m; stems 3–4 m high: 31187; W 2782. "Caballo chupa"

LYCOPODIACEAE (3) (* det. B. Øllgaard; ** det. A. R. Smith)
   Ref.: Øllgaard, B. 1988. Fl. Ecuador 33: 1–155.

1. *Huperzia acerosa* (Sw.) Holub
   Sec. for., 1200–1400 m; ep.: Filskov 37093*.

2. *Huperzia acifolia* (Rolleri) Rolleri & Deferrari
   Sec. for., 1900–2000 m: L 14040*; Øllgaard 100623.

3. *Huperzia curvifolia* (Kunze) Holub
   Sec. for., 1200–1500 m; ep.: 27051**; V 12241**; Sodiro s.n., 1874.

4. *Huperzia eversa* (Poir.) B. Øllg.
   Sec. for., Cerro Negro, 1900 m; terr.: 30456.

5. *Huperzia hippuridea* (H. Christ) Holub
   Upper mont. for., 2750–2800 m; terr., stems erect: 31513.

6. *Huperzia homocarpa* (Herter) Holub
   Sec. for., 1200–1500 m; ep.: Sodiro s.n., Nanegal.

7. *Huperzia linifolia* (L.) Trevis.
   Sec. for., 1150–1600 m; ep.: 27018, 27050, 28113*, 28266*, 29290, 29474, 29554; C 12422; V 12249*; W 2678, 2755; Filskov 37113; Sparre 14887.

8. *Huperzia myrsinites* (Lam.) Trevis.
   Sec. for., 1400–1700 m; ep.: M 5241*.

9. *Huperzia reflexa* (Lam.) Trevis.
   Mossy banks, 1150–2500 m; terr.: 27227, 28819, 31148, 32968; L 13640; W 2618**.

10. *Huperzia taxifolia* (Sw.) Trevis.
    Sec. for., 1200–1500 m; ep., stems pendent: V 12257 (det. R. Moran).

11. *Huperzia* cf. *unguiculata* B. Øllg.
    Mossy banks, 1200–1500 m; terr.: 27856*, 28152.

12. *Huperzia wilsonii* (Underw. & F. E. Lloyd) B. Øllg.
    Prim. & sec. for., 1200–1800 m; terr.: 28217*, 29319, 30316; Filskov 37031*, 37091*.

13. *Lycopodiella cernua* (L.) Pic. Serm.
    Prim. & sec. for., 1600–2700 m: 27049, 27100, 29469, 30564.

14. *Lycopodium clavatum* L. ssp. *clavatum*
Prim. & sec. for., mossy banks, 1200–2750 m: 27231, 27298, 27787*, 29495, 30469; L 14041; Haught 3185.

15. *Lycopodium thyoides* Humb. & Bonpl. ex Willd.
Upper montane for., Cerro Montecristi, 2500–2750 m: 29526; L 13666.

SELAGINELLACEAE (1) (* det. I. Valdespino; ** det. R. Moran; *** det. A. R. Smith)
Ref.: Alston, A. H. G., A. C. Jermy & J. M. Rankin. 1981. Bull. Brit. Mus. (Nat. Hist.) Bot. 9: 233–330.

1. *Selaginella diffusa* Spring
Sec. for., 1350–2050 m; prostrate: 27211*, 27268*, 28027*, 30320, 30482, 31921.

2. *Selaginella geniculata* (C. Presl) Spring
Sec. for., 1200–1500 m: 27032, 27219*, 27793, 27919*, 29063*; W 2699; Z 87.

3. *Selaginella kunzeana* A. Braun
Sec. for., 1250 m; forming mats: 27026 ex p.*; Filskov 37098.

4. *Selaginella lingulata* Spring
Prim. & trans. for., 1250–2375 m: 27020*, 27026 ex p.*, 27099*, 27860*, 29467*; W 2601***.

5. *Selaginella macilenta* Baker
Sec. for., 1250–1300 m; forming mats: 27923*.

6. *Selaginella novae-hollandiae* (Sw.) Spring
Sec. for., 1200–2200 m; stems erect: 27228*, 27267*, 27800*, 27926*; W 2616***; 1917, Mille s.n.

7. *Selaginella sericea* A. Braun
Prim. & sec. for., 1500–1900 m: 27805**, 27829**, 28334**, 29218; F 1527; M 5227*; W 2597.

8. *Selaginella trisulcata* Aspl.
Sec. for., Cerro Santa Lucia, 1700 m: 28347**.

9. *Selaginella* sp. 1
   Rip. for., 1350–1400 m: 30287.

ASPLENIACEAE (1) (* det. A. R. Smith; ** det. R. Moran)
Ref.: Stolze, R. G. 1986. Fl. Ecuador 23: 1–83.

1. *Asplenium alatum* Humb. & Bonpl. ex Willd.
   Prim. for., 1900–2000 m: 27208*; W 2607.

2. *Asplenium auriculatum* Sw.
   Prim. for., 1900–2750 m; ep.: 28868*, 29435*; W 2619*, 2781*.

3. *Asplenium auritum* Sw.
   Prim. & sec. for., 1200–2250 m; ep.: 27055*, 27457*, 27574*, 30302*;
   Filskov 37104, 37115 (both det. R. G. Stolze).

4. *Asplenium cristatum* Lam.
   Rip. for., 1900 m; terr.: W 2605A.

5. *Asplenium cuspidatum* Lam.
   Prim. & sec. for., 1200–2750 m; usually ep.: 27577*, 27578, 27605*, 27726*,
   27818*, 27858*, 29140*, 29280*, 29332*, 29535*, 30020*; W 2687;
   Filskov 37097, 37105.

6. *Asplenium flabellulatum* Kunze
   Prim. for., 1900–2250 m; ep.: 28871*, 29144*; W 2596*, 2611*.

7. *Asplenium fragrans* Sw.
   Sec. & upper mont. for., 1250–2750 m; ep.: W 2694*, 2762, 2780.

8. *Asplenium harpeodes* Kunze
   Sec. upper mont. for., 2700–2750 m; ep.: W 2778.

9. *Asplenium juglandifolium* Lam.
   Sec. for., 1300–1800 m; ep.: 29124*; C 12400**, 12410 (det. B. Øllgaard).

10. *Asplenium maxonii* Lellinger
    Prim. for., 1800 m; ep.: 27448*.

11. *Asplenium myriophyllum* (Sw.) C. Presl
    Sec. for., 1800–2000 m: Øllgaard 99759**.

12. *Asplenium pteropus* Kaulf.
Sec. for., 1250–1600 m; ep.: 27479*, 29323*, 30271*, 30507*; M 5238; W 2635, 2697*, 2701*.

13. *Asplenium purpurascens* Mett. ex Kuhn
Sec. for., 1250 m; ep.: 27727*; Sodiro asp8.74 [*Asplenium melanopus* Sodiro].

14. *Asplenium rosenstockianum* Brade
Sec. for., 1200–1300 m: 27777*. [Not in CVPE.]

15. *Asplenium rutaceum* (Willd.) Mett.
Prim. for., 1400–2000 m; ep. or terr.: 27823*, 28239*; M 5207.

16. *Asplenium serra* Langsd. & Fisch.
Prim. for., 1700 m; ep., fronds nearly 1 m: 28346*.

17. *Asplenium serratum* L.
Sec. for., 1250–1300 m; ep., leaves undivided: W 2636.

18. *Asplenium* aff. *tricholepis* Rosenst.
Sec. for., 1350 m; terr.: 32363.

19. *Asplenium uniseriale* Raddi
Prim. for., 1800–1900 m: 28772*.

AZOLLACEAE (1) (* det. A. R. Smith)

1. *Azolla filiculoides* Lam. ?
Seep along road to Santa Marianitas, 1250 m: 27380*.

BLECHNACEAE (1) (* det. A. R. Smith; ** det. R. Moran)

1. *Blechnum cordatum* (Desv.) Hieron.
Sec. for., 1200–1600 m; fronds 1.5 m: 27019*, 27342*, 27641*; W 2622*.

2. *Blechnum divergens* (Kunze) Mett.
Prim. for., 1900–2175 m; ep. or terr.: 28875*, 30187*; Croat 72869**, 72875**.

3. *Blechnum ensiforme* (Liebm.) C. Chr.
Prim. for., 1700–2000 m; ep.: 27108*; C 5907**, 39700. [Collections of this species have been referred to *B. binervatum* (Poir.) C. V. Morton & Lellinger.]

4. *Blechnum glandulosum* Link
Sec. for., 1200–1350 m; ep. or terr: 27047*, 27523*, 28671*; W 2626, 2689.

5. *Blechnum lherminieri* (Bory) Mett.
Prim. for., 1750–2300 m; ep. or terr.: 27109*, 28291*, 29223*.

6. *Blechnum occidentale* L.
Prim. & sec. for., 1250–2300 m; ep. or terr.: 27031*, 27244*, 28012*, 28916*; M 5215; T 189**.

7. *Blechnum stipitellatum* (Sodiro) C. Chr.
Prim. for., 2200 m: 29146*.

8. *Blechnum wardiae* Mickel & Beitel
Sec. for., 1400–1700 m: M 5226.

9. *Blechnum* aff. *divergens* (Kunze) Mett.
Prim. for., 2275–2300 m: 29135*; F 1426.

CYATHEACEAE (3) (* det. A. R. Smith; ** det. R. Moran)
Refs.: Tryon, R. M., & R. G. Stolze. 1989. Fieldiana Bot. II. 20 (Pteridophyta of Peru, I(1): 111–139. Moran, R. C., & B. Øllgard. 1998. Nordic J. Bot. 18: 431–439.

1. *Alsophila cuspidata* (Kunze) R. M. Tryon
Loma Pahuamba, 2300 m: F 1447.

2. *Alsophila erinacea* (H. Karst.) D. S. Conant
Sec. for., 1500–1600 m; trunk 2.5 m, fronds 3–4 m: 27497*; G 73220**.

3. *Alsophila incana* (H. Karst.) D. S. Conant
Sec. for., 1200–1900 m; trunk to 15 m, fronds 1–2.3 m: 28971*; F 1005; W 2628*, 2756.

4. *Cyathea brunnescens* (Barrington) R. C. Moran
Sec. for., El Carmen, 1200 m; trunk 3 m: 29410*.

5. *Cyathea caracasana* (Klotzsch) Domin
    Prim. & sec. for., 1250–2750 m; trunk to 8–12 m, fronds 2–2.5 m: 28000*, 28778*, 29083*, 29356*, 30172*, 30554*; C 39764; G 69914**, 73133**, 73234**; M 5246; Q 16**; W 2758, 2777. "Helecho pelado", "Helecho paramal"

6. *Cyathea halonata* R. C. Moran & B. Øllg.
    Prim. & sec. for., 1400–1750 m; trunk to 2 m, fronds to 2 m: 30028; G 73187, 73245 [paratype]; M 5257. [Not in CVPE.]

7. *Cyathea multiflora* Sm.
    Sec. for., 1300–1650 m; trunk 1.5 m: 27494*, 29299*; G 69946**.

8. *Cyathea nigripes* (C. Chr.) Domin
    Sec. for., 1250–1300 m; trunk 1.5 m, fronds 3 m: W 2633*.

9. *Cyathea pilosissima* (Baker) Domin
    Sec. for., 1250 m; trunk 2.5 m, fronds 3 m: 27024*.

10. *Cyathea poeppigii* (Hook.) Domin
    Sec. for., 1200–1250 m; trunk 1 m, fronds 2 m: 28304*.

    [W 2632, 2770, 2771 represent one or more additional species of *Cyathea*.]

11. *Sphaeropteris quindiuensis* (H. Karst.) Tryon
    Sec. for., 1400–1700 m: G 73133A**; M 5260.

DENNSTAEDTIACEAE (4) (* det. A. R. Smith)
    Ref.: Tryon, R. M., & R. G. Stolze. 1989. Fieldiana Bot. II. 22 (Pteridophyta of Peru II, 13): 94–122.

1. *Dennstaedtia arborescens* (Willd.) Ekman ex Maxon
    Prim. for., 1350–2500 m; fronds 1.5–2.5 m: M 5258; V 12339 (det. R. Moran); W 2650, 2705. [Not in CVPE.]

2. *Dennstaedtia dissecta* (Sw.) T. Moore
    Prim. for., 2000 m; ep.: 30144*.

3. *Dennstaedtia sprucei* T. Moore
    Rip. for., 1900 m; terr., fronds 2.5 m: W 2602*.

4. *Hypolepis steubelii* Hieron.
    Sec. for., 1400–1900 m; fronds 2 m: 30054*; M 5244; W 2617.

5. *Pteridium arachnoideum* (Kaulf.) Maxon
    Sec. & upper mont. for., 1600–2700 m: 30065*; W 2779. "El Asa"

6. *Saccoloma domingense* (Spreng.) C. Chr.
    Sec. for., 1250–1500 m; trunk 1 m, fronds 2–3 m: 28740*, 29287*; W 2629, 2674.

DICKSONIACEAE (1)
    Ref.: Tryon, R. 1986. Fl. Ecuador 27: 3–6.

1. *Dicksonia sellowiana* Hook.
    Prim. for., 2400–2500 m; stems to 15 m: F 1331.

DRYOPTERIDACEAE (12) (* det. A. R. Smith; ** det. R. Moran; *** det. J. Mickel)
    Ref.: Tryon, R. M., & R. G. Stolze. 1991. Fieldiana Bot. II. 27 (Pteridophyta of Peru IV, 17): 1–176 (*Elaphoglossum* by J. Mickel).

1. *Ctenitis sloanei* (Spreng.) C. V. Morton
    Sec. for., 1300 m: M 5188.

2. *Didymochlaena truncatula* (Sw.) J. Sm.
    Sec. for., 1350–1600 m; fronds 1.5 m: 27474*, 29981*.

3. *Diplazium ambiguum* Raddi var. *ambiguum*
    Sec. for., 1250–1550 m; fronds 2 m: 27046*; Stolze & Stolze 1745.

3A. *Diplazium ambiguum* Raddi var. *dissectum* Stolze
    Collected with var. *ambiguum* 1 km south of Nanegalito: Stolze & Stolze 1752, 1754.

4. *Diplazium centripetale* (Baker) Maxon, vel aff.
    Rip. for., 1250–1300 m: W 2679.

5. *Diplazium chimborazense* (Spruce ex Baker) H. Christ
    Prim. & sec. for., 1350–1800 m; ep., fronds 2.5–3 m: 28195, 29989*; M 5225; Pacheco 3288, 3289, 3290.

6. *Diplazium costale* (Sw.) C. Presl var. *robustum* (Sodiro) Stolze
   Prim. for., 2000–2300 m; ep. or terr., to 2 m: 30152*; F 1436.

7. *Diplazium diplazioides* (Klotzsch & H. Karst.) Alston
   Sec. for., 1560 m.: Stolze & Stolze 1751.

8. *Diplazium divississimum* (Baker) H. Christ
   Prim. & sec. for., 1200–2000 m; fronds 1–2 m: 27266*, 28193*, 28237*,
   28777*, 30276*, 31104*; M 5203; W 2710; Sodiro s.n.

9. *Diplazium expansum* Willd.
   Sec. for., 1200 m; rhizome erect: V 12290**.

10. *Diplazium ferulaceum* (T. Moore ex Hook.) Lellinger
    Sec. for., 1300–2000 m; terr.: 27209*, 28171*; T 183; V 12331; W 2605;
    Stolze & Stolze 1748.

11. *Diplazium* cf. *hians* Klotzsch
    Prim. for., 2500 m; fronds 2 m: V 12338**.

12. *Diplazium lindbergii* (Mett.) H. Christ
    Prim. & sec. for., 1275–1800 m; fronds 2.5 m: 28207*, 30281*; W 2704*.

13. *Diplazium lonchophyllum* Kunze
    Sec. for., 1250–1560 m: 27771*; Stolze & Stolze 1747.

14. *Diplazium macrophyllum* Desv.
    Sec. for., 1250–1560 m; fronds 1.5–3 m: 27212*, 27942*, 28179**, 29271;
    Filskov 37117; Stolze & Stolze 1753.

15. *Diplazium moccenianum* (Sodiro) C. Chr.
    Sec. for., 1200–1250 m: 28287*.

16. *Diplazium rivale* (Baker) Diels
    Sec. for., 1400–1700 m: M 5252.

17. *Diplazium roemerianum* (Kunze) C. Presl var. *roemerianum*
    Sec. for., Inca trail, 1500–1700 m; ep.: Q 128**; 1 km south of Nanegalito,
    1560 m: Stolze & Stolze 1749.

18. *Diplazium sanderi* (C. Chr.) Pacheco
    Nanegal: Sodiro s.n.

19. *Elaphoglossum albescens* (Sodiro) H. Christ
    Prim. for., 1500–1650 m; ep.: 27506*.

20. *Elaphoglossum auricomum* (Kunze) T. Moore
    Sec. for., 1500–1650 m; ep.: 27508***.

21. *Elaphoglossum bakeri* (Sodiro) H. Christ
    Prim. & rip. for., 1500–2000 m; fronds to 1.5 m: 28021***, 28244*, 30369*, 32436*; C 12408; M 5233*; W 2608.

22. *Elaphoglossum cuspidatum* (Willd.) T. Moore
    Sec. for., 1250–1300 m; ep.: M 5202*; W 2693*.

23. *Elaphoglossum erinaceum* (Fée) T. Moore
    Sec. for., 1400–1700 m; ep.: M 5212*.

24. *Elaphoglossum eximium* (Mett.) H. Christ
    Prim. & sec. for., 1300–2000 m; ep. or epipetric: 28717***, 28854***, 32456*.

25. *Elaphoglossum latifolium* (Sw.) Sm.
    Sec. upper mont. for., 2700–2750 m; ep.: W 2773.

26. *Elaphoglossum lloense* (Hook.) Moore
    Sec. upper mont. for., 2700–2750 m; ep.: W 2772.

27. *Elaphoglossum oblanceolatum* C. Chr.
    Sec. for., 1250–1350 m; ep.: 28794*; W 2698.

28. *Elaphoglossum orbignyanum* (Fée) T. Moore
    Sec. for., 1250 m; ep.: 27524*.

29. *Elaphoglossum* aff. *paleaceum* (Hook. & Grev.) Sledge
    Sec. for., 1300–1500 m; ep.: 28133***, 30004***; W 2690.

30. *Elaphoglossum papillosum* (Baker) Christ
    Prim. for., 2100 m; ep.: P 3593*.

31. *Elaphoglossum peltatum* (Sw.) Urb.
    Prim. & sec. for., 1200–2400 m; ep.: 27603, 28018*, 28867, 29515; W 2737.
    [Often treated as *Peltapteris peltata* (Sw.) C. V. Morton]

32. *Elaphoglossum phoras* Mickel
    Sec. for., 1200–1700 m; ep.: M 5231. [Not in CVPE.]

33. *Elaphoglossum pseudoboryanum* Mickel
    Sec. for., 1200–1700 m; ep.: M 5232.

34. *Elaphoglossum pygmaeum* (Mett. ex Kuhn) H. Christ
    Sec. for., 1300–1700 m; ep.: 27979*, 30300*; M 5521*; V 12313*, 12325*;
    W 2741.

35. *Elaphoglossum tectum* (Humb. & Bonpl. ex Willd.) T. Moore
    Prim. for., 1700–1800 m: 27859*.

36. *Elaphoglossum urbanii* (Sodiro) C. Chr.
    Prim. & sec. for., 1250–2000 m; ep.: 27052***, 28234A***.

[Several additional collections of *Elaphoglossum* remain unidentified to species.]

37. *Hemidictyum marginatum* (L.) C. Presl
    Sec. for., 1300–1350 m; terr., fronds 3 m: W 2706.

38. *Lastreopsis effusa* (Sw.) Tindale
    Sec. for., Río Alambi, 1100–1200 m; fronds 2 m: 29267*; V 12291*.

39. *Megalastrum andicola* (C. Chr.) A. R. Smith & R. C. Moran
    Sec. for., 1300–2200 m: F 1480; M 5200, 5220; P 3600**; V 12324**.

40. *Megalastrum biseriale* (Baker) A. R. Smith & R. C. Moran
    Sec. for., 1400–1700 m: M 5229.

[*Megalastrum subincisum* (Willd.) A. R. Smith & R. C. Moran was collected by Cerón (1483*, 1489*) just east of our boundary.]

41. *Megalastrum vastum* (Kunze) A. R. Sm.
    Sec. for., 1450–1500 m; terr.: W 2709.

42. *Nephrolepis cordifolia* (L.) C. Presl
Loma Pahuamba, 1750 m: F 1506.

43. *Nephrolepis pectinata* (Willd.) Schott
Sec. for., 1200–1900 m; ep.: 27029, 27528*, 28016*, 28790*, 28872*; W 2708*.

44. *Nephrolepis pendula* (Raddi) J. Sm.
Prim. & sec. for., 1500–1750 m; ep.: 27460*, 27344*.

45. *Polybotrya altescandens* C. Chr.
Sec. for., 1250–2000 m; terr. or ep., fronds to 2 m: 27213*, 27272*, 27576*, 29312*, 30192; W 2582, 2675.

46. *Polybotrya polybotryoides* (Baker) H. Christ
Prim. for., 1550–2000 m; ep. climber: 28246*; G 73165**.

47. *Polystichum platyphyllum* (Willd.) C. Presl
Banks in rip. for., 1900–2000 m; terr., fronds c. 1 m: 31836; W 2610*.

48. *Stigmatopteris ichthiosma* (Sodiro) C. Chr.
Sec. for., 1400–1900 m; fronds 2.5 m: M 5250; W 2606.

49. *Tectaria antioquiana* (Baker) C. Chr.
Sec. for., 1250–1350 m; fronds 1.5 m: 29990*; W 2682.

50. *Tectaria chimborazensis* C. Chr. ?
Sec. for., 1200 m: 28142*.

51. *Tectaria lizarzaburui* (Sodiro) C. Chr.
Prim. for., 1300–2030 m: 30204*; V 12333*; W 2615.

GLEICHENIACEAE (3) (* det. A. R. Smith)
Ref.: Tryon, R. M., & R. G. Stolze. 1989. Fieldiana Bot. II. 20 (Pteridophyta of Peru I, 1): 37–49.

1. *Diplopterygium bancroftii* (Hook.) A. R. Sm.
Sec. for., banks, 1200–1700 m; 1–2 m high: 30440*; W 2684, 2747; Nanegal, 1872, 1874, Sodiro s.n.

2. *Gleicheniella pectinata* (Willd.) Ching [*Dicranopteris pectinata* (Willd.) Underw.]
Sec. for., 1250–1300 m; scrambling on banks: 31086*; W 2624; Østergaard & Andersen 10818.

3. *Sticherus bifidus* (Willd.) Ching
Sec. for., 1200–1600 m; scrambling, fronds to 4 m: 27030*, 27343*, 27874*, 28681*; W 2625, 2731; 1874, Sodiro s.n.; Sparre 14879.

4. *Sticherus blepharolepis* (Sodiro) Ching
Sec. for., 1350–1700 m: M 5196, 5248; Øllgaard 899; Ostergaard & Andersen 10811, 10817. "Perfilillo del cerro"

5. *Sticherus hypoleucus* (Sodiro) Copeland
Sec. for., Nanegalito to Nanegal: Ostergaard & Andersen 10822; 1872, Sodiro s.n.

[*Sticherus revolutus* (Humb. & Bonpl. ex Kunth) Ching, from between Alaspungo & Gualea, Heilborn s.n., Holmgren 711, is probably at an elevation above our limits.]

[*Sticherus rubiginosus* (Mett.) Nakai, Østergaard & Andersen 10732, 10814, is apparently extralimital.]

6. *Sticherus tomentosus* (Cav. ex Sw.) A. R. Sm.
Prim. for., 1550–1675 m: 28701*, 30005*.

GRAMMITIDACEAE (5) (* det. A. R. Smith)
Ref.: Morton, C. V. 1967. Contr. U. S. Nat. Herb. 38: 85–123. 1971. Phytologia 22: 71–82.

1. *Cochlidium serrulatum* (Sw.) L. E. Bishop
Sec. for., 1250–1500 m; ep.: 27903*, 27920*; W 2752.

2. *Enterosora trifurcata* (L.) L. E. Bishop
Prim. for., Loma Cachillacta, 1700 m; ep.: 30439*; Nanegal, Jameson.

3. *Lellingeria suspensa* (L.) A. R. Sm. & R. C. Moran
Sec. for., 1300–1500 m; ep.: 27902*; W 2751.

4. *Melpomene anfractuosa* (Kunze ex Klotzsch) A. R. Sm. & R. C. Moran
   Prim. & sec. for., 1250–1900 m; ep.: 27461*, 27573*, 27763*, 27925*, 30002*, 30299*; M 5218; W 2594, 2677, 2696.

5. *Melpomene assurgens* (Maxon) A. R. Sm. & R. C. Moran
   Prim. for., 1700–2500 m; ep.: 27444*, 27464*, 27753*, 27816* 28228*, 28727*, 29421*, 30585*; W 2593*, 2783*.

6. *Melpomene flabelliformis* (Poir) A. R. Sm. & R. C. Moran
   Prim. for., 2000 m; ep.: C 5909 (det. P. M. Jørgensen).

7. *Melpomene melanosticta* (Kunze) A. R. Sm. & R. C. Moran
   Prim. for., 1750 m; ep.: 27106*.

8. *Melpomene moniliformis* (Lag. ex Sw.) A. R. Sm. & R. C. Moran
   Sec. for., 1400–2000 m; ep.: M 5213, 5242; Q 132.

9. *Terpsichore alsopteris* (C. Morton) A. R. Sm.
   Prim. & sec. for., 1400–2500 m; ep.: 30213*; L 13669*; M 5236; P 3583*; V 12323; W 2609*.

10. *Terpsichore cultrata* (Borg ex Willd.) A. R. Sm.
    Sec. for., 1400–1700 m; ep.: M 5239, 5255*.

11. *Terpsichore lanigera* (Desv.) A. R. Sm.
    Prim. for., 2000 m; ep.: 28238*.

12. *Terpsichore lehmanniana* (Hieron.) A. R. Sm.
    Sec. for., 1350 m; ep.: 29992*.

13. *Terpsichore taxifolia* (L.) A. R. Sm.
    Prim. & sec. for., 1300–1750 m; ep.: 27692*, 28226*; M 5222, 5230.

14. *Terpsichore* sp. 1 (ined.)
    Sec. for., 1900–2000 m; ep.: W 2609 ex p*.

HYMENOPHYLLACEAE (2) (* det. A. R. Smith)
Ref.: Tryon, R. M., & R. G. Stolze. 1989. Fieldiana Bot. II. 20 (Pteridophyta of Peru I, 1): 49–98.

1. *Hymenophyllum dependens* (Sw.) Sw. (vel aff.)
   Sec. for., 1200–2725 m; ep.: 27604*; W 2637*, 2766*.

2. *Hymenophyllum fragile* (Hedw.) C. V. Morton (vel aff.)
   Sec. for., 1400–1700 m; ep.: M 5235.

3. *Hymenophyllum fucoides* (Sw.) Sw.
   Prim. for., 1400–2450 m; common ep.: 27104*, 27456*, 27819*, 28215*, 29225*; F 1377; M 5237.

4. *Hymenophyllum jamesonii* Hook.
   Upper mont. for., Cerro Montecristi; 2750 m; ep.: H 29541.

5. *Hymenophyllum microcarpum* Desv.
   Prim. & sec. for., 1350–1750 m; ep. or terr.: 27765*, 30296*, 30430*; M 5234, 5245.

6. *Hymenophyllum myriocarpum* Hook.
   Sec. & upper mont. for., 2000–2725 m; ep.: 28026*, 29508*.

7. *Hymenophyllum plumieri* Hook. & Grev.
   Prim. for., 2250–2725 m; ep.: 28914*, 29479*; M 5240*; W 2767*.

8. *Hymenophyllum polyanthos* (Sw.) Sw.
   Sec. for., 1250–1300 m; ep.: W 2638*.

9. *Hymenophyllum ruizianum* (Klotzsch) Kunze
   Sec. for., Cerro Negro, 1500–2000 m; ep.: 28025*.

10. *Hymenophyllum trichomanoides* Bosch
    Sec. upper mont. for., 2700–2750 m; ep.: W 2763*.

11. *Hymenophyllum trichophyllum* Kunth
    Sec. upper mont. for., 2700–2750 m; ep.: W 2768.

12. *Trichomanes capillaceum* L.
    Prim. for., 2250–2300 m; ep.: 28915*.

13. *Trichomanes lucens* Sw.
Upper mont. for., 2575 m; ep.: 29480*.

14. *Trichomanes radicans* Sw.
Rip. for., on banks, 2000 m; terr., fronds from a creeping rhizome: 32430.

15. *Trichomanes reptans* Sw.
Sec. for., 1300–1400 m; ep. or terr.: 28957*, 30278*.

16. *Trichomanes rigidum* Sw.
Sec. for., 1200–1350 m; terr.: 29306*; W 2700.

LOPHOSORIACEAE (1) (* det. R. Moran)
Ref.: Tryon, R. M., & R. G. Stolze. 1989. Fieldiana II. 20 (Pteridophyta of Peru I, 1): 107–109.

1. *Lophosoria quadripinnata* (J. F. Gmel.) C. Chr.
Sec. for., 1250–1700 m; trunk 1.5–2.5 m, fronds 2.5–5 m: 27820*; W 2627.

MARATTIACEAE (1) (* det. H. Tuomisto & R. Moran)
Ref.: Tryon, R. M., & R. G. Stolze. 1989. Fieldiana II. 20 (Pteridophyta of Peru I, 1): 13–20. Tuomisto, H., & R. Moran. 2000(?). Fl. Ecuador (in press).

1. *Danaea erecta* H. Tuomisto & R. C. Moran
Prim. & sec. for., 1200–1750 m; fronds to 2 m: 27496*, 27107; M 5256*, 5950; W 2683; Øllgaard 908*; Stolze & Stolze 1750*. [Some specimens have been named *D. moritziana* C. Presl, but that species has not been verified for our area.]

OPHIOGLOSSACEAE (1)
Ref.: Tryon, R. M., & R. G. Stolze. 1989. Fieldiana, Bot. II. 20 (Pteridophyta of Peru I, 1): 5–13.

1. *Ophioglossum reticulatum* L.
Banks, 1300–1350 m: M 5189, 5201; Z 191.

POLYPODIACEAE (5) (* det. A. R. Smith)

1. *Campyloneurum amphostemon* (Kunze ex Klotzsch) Fée ?
Prim. & sec. for., 1250–2725 m; ep.: 27120*; W 2642, 2774*.

2. *Campyloneurum angustifolium* (Sw.) Fée
   Sec. for., 1250–1350 m; ep.: 27023, 28136A*; Addison F70*.

3. *Campyloneurum aphanophlebium* (Kunze) T. Moore
   Sec. for., 1250–1300 m; ep.: W 2634*, 2639*, 2686*.

4. *Campyloneurum brevifolium* (Link) Link
   Sec. for., 1200 m; ep.: 27525*.

5. *Campyloneurum magnificum* T. Moore
   Sec. for., 1200 m; ep., fronds to 1.3 m: 27527*.

6. *Campyloneurum ophiocaulon* (Kunze ex Klotzsch) Fée
   Prim. & sec. for., 1750–2000 m; ep.: 27114*, 28770*, 30171*, 30174*, 32419*; W 2600.

7. *Campyloneurum phyllitidis* (L.) C. Presl
   Sec. for., 1300–1900 m; ep.: 27897*; F 1059. "Calaguala"

8. *Campyloneurum repens* (Aubl.) C. Presl
   Sec. for., 1200 m; ep.: 27770*.

9. *Campyloneurum sphenodes* (Kunze ex Klotzsch) Fée
   Prim. for., 2075 m; ep.: 29247*.

10. *Microgramma piloselloides* (L.) Copel.
    Prim. & sec. for., 1200–1800 m; ep.: 27606*, 27861*, 28136, 29331*; M 5191; W 2681; Filskov 31707.

11. *Microgramma reptans* (Cav.) A. R. Sm.
    Sec. for., 1250–1300 m; ep.: 27027*; W 2630.

12. *Microgramma tecta* (Kaulf.) Alston
    Sec. for., 1150–1600 m; ep.: 27345*, 27572*, 31269*; M 5191A; W 2643.

13. *Niphidium crassifolium* (L.) Lellinger
    Prim. & sec. for., 1200–2000 m; common ep.: 27216*, 27463*, 28242*, 28295*, 28969*; Filskov 37108. "Calaguala"

14. *Pecluma divaricata* (E. Fourn.) Mickel & Beitel
    Sec. for., 1250–1700 m; ep.: 30510*, 32913*, 32914*; Q 136 (det. R. Moran); W 2703*.

15. *Pecluma eurybasis* (C. Chr.) M. G. Price
    Prim. & sec. for., 1300–2750 m; common ep.: 27025*, 27110*, 28182*, 28952*, 29122, 29296*, 30275*, 30553*; W 2644, 2775.

16. *Pecluma ptilodon* (Kunze) M. G. Price
    Sec. for., 1200–1700 m; ep.: 27346*, 27520*; M 5210, 5214.

17. *Polypodium adnatum* Kunze ex Klotzsch
    Prim. & sec. for., 1400–2000 m; ep., fronds to 1 m: 28877*; M 5219.

18. *Polypodium caceresii* Sodiro
    Prim. & sec. for., 1700–1900 m; ep.: 28057*, 28190*; W 2591.

19. *Polypodium dasypleuron* Kunze
    Sec. for., 1250–1300 m: W 2645.

20. *Polypodium dissimile* L.
    Sec. for., 1200–1300 m; ep.: 27640*; M 5194.

21. *Polypodium fayorum* R. C. Moran & B. Øllgaard
    Prim. & sec. for., 1200–1900 m; ep.: 27529*, 27939*, 28869*, 29102*; M 5190; W 2707*, 2757*; Filskov 37106** [paratype]. [Not in CVPE.]

22. *Polypodium fraxinifolium* Jacq.
    Prim. & sec. for., 1400–2000 m; ep.: 27217*, 27218*, 28031*, 28235*, 28716*, 30167*, 30173*; W 2595, 2613; Hurtado 1440.

23. *Polypodium levigatum* Cav.
    Prim. & sec. for., 1250–2300 m; ep.: 27054*, 27271*, 27526, 27956*, 29295; W 2753; Hurtado 1423.

24. *Polypodium monosorum* Desv. ?
    Upper mont. for., 2700–2750 m; ep.: 30527*.

25. *Polypodium patentissimum* Mett. ex Kuhn
    Sec. for., 1400–1700 m: M 5228.

26. *Polypodium pseudoaureum* Cav.
    Sec. for., 1250–1350 m; ep.: 27458*, 28734*, 32915.   [*Phlebodium pseudoaureum* (Cav.) Lellinger in CVPE.]

27. *Polypodium segregatum* Baker
    Sec. for., banks, 2300 m: W 2759*.

28. *Polypodium* cf. *sporadolepis* Kunze ex Mett.
    Sec. for., 1250–1350 m; ep.: 27033*, 27053*. [Not in CVPE.]

29. *Polypodium thysanolepis* A. Braun ex Klotzsch
    Sec. for., banks, 2300 m: W 2786.

30. *Polypodium (Pleopeltis)* sp. 1
    Sec. for., 1200–1350 m; ep.: 27459*, 27521*.

PTERIDACEAE (5) (* det. A. R. Smith)
    Ref.: Tryon, R. M., & R. G. Stolze. 1989. Fieldiana Bot. II. 22 (Pteridophyta of Peru II, 13): 2–83.

1. *Adiantum concinnum* Humb. & Bonpl. ex Willd.
    Banks & dist. for., 1150–1400 m: 27269*, 27370, 28050, 28298, 28820*; W 2685. "Culantrillo"

2. *Adiantum macrophyllum* Sw.
    Prim. & sec. for., 1250–1600 m: 27168*, 27472; W 2631.

3. *Adiantum patens* Willd.
    Sec. for., 1400–1950 m: 27270*; W 2612.

4. *Adiantum poiretii* Wikstr.
    Sec. for., banks, 2000 m: W 2784.

5. *Adiantum urophyllum* Hook.
    Sec. for., 1250–1300 m: 27940*; V 12334 (det. R. Moran).

6. *Cheilanthes myriophylla* Desv.
    Sec. for., banks, 2300 m: W 2760.

7. *Pellaea ternifolia* (Cav.) Link
    Sec. for., banks, 2300 m: W 2785.

8. *Pityrogramma calomelanos* (L.) Link
   Banks, 1250–1350 m: 27048. "Gallito", "Shapumbilla"

9. *Pityrogramma ebenea* (L.) Proctor
   Banks, 1700–2000 m; fronds to 2.5 m long: 28189*, 30220*, 32418; W 2598.

10. *Pityrogramma trifoliata* (L.) Tryon
    Sec. for., 1200 m: 27607*.

11. *Pteris altissima* Poir
    Sec. for., 1200 m; 1.5 m tall: V 12292*.

12. *Pteris* cf. *consanguinea* Mett. ex Kuhn
    Sec. for., 1400–1700 m: M 5251*.

13. *Pteris fraseri* Mett. ex Kuhn
    Sec. for., 1250–1350 m: 29997*; W 2680*.

14. *Pteris muricatopedata* Arbeláez
    Sec. for., 1300 m; 3 m tall: V 12332 (det. A. L. Arbeláez).

15. *Pteris podophylla* Sw.
    Sec. for. & river banks, 1400–1600 m: 27220*, 27462*.

16. *Pteris quadriaurita* Retz.
    Sec. for., 1200–1550 m: 29320*, 30006*; V 12287, 12329; Filskov 37100 (last three det. A. L. Arbeláez).

THELYPTERIDACEAE (2) (* det. A. R. Smith)
   Ref.: Smith, A. R. 1983. Fl. Ecuador 18: 1–147.

1. Macrothelypteris torresiana (Gaud.) Ching
   Sec. for., 1250–1600 m: 27223*, 27224*, 27225*, 30414*; W 2620. [The only introduced pteridophyte in our flora.]

2. *Thelypteris amphioxipteris* (Sodiro) A. R. Sm.
   Sec. for., 1200–1400 m: 27222*, 29062*; Filskov 37099*.

3. *Thelypteris andreana* (Sodiro) C. V. Morton
   Sec. for., 1400–1700 m: M 5195, 5243.

4. *Thelypteris cheilanthoides* (Kunze) Proctor
　　Sec. for., 1200–1250 m: 28683*; W 2621*.

5. *Thelypteris cinerea* (Sodiro) A. R. Sm.
　　Sec. for., 1250–1300 m; ep.: 27728*, 27938*.

6. *Thelypteris concinna* (Willd.) Ching
　　Sec. for., 1200–1400 m: 27118*; Filskov 37118*.

7. *Thelypteris dentata* (Forssk.) E. P. St. John
　　Sec. for., 1200–1300 m: 27221*, 28107*.

8. *Thelypteris euchlora* (Sodiro) A. R. Sm.
　　Prim. for., 2000 m: 30137*.

9. *Thelypteris hispidula* (Decne.) C. F. Reed
　　Sec. for., 1200–1250 m: 27117*.

10. *Thelypteris linkiana* (C. Presl) Tryon
　　Sec. for., 1200–1700 m: 27119*; M 5198; V 12288*.

11. *Thelypteris oligocarpa* (Humb. & Bonpl. ex Willd.) Ching
　　Sec. for., 1400–1700 m: M 5197, 5247*.

12. *Thelypteris opposita* (Vahl) Ching
　　Sec. for., 1200–1250 m: W 2623*.

13. *Thelypteris pachyrhachis* (Kunze ex Mett.) Ching
　　Sec. for., 1300 m: V 12327*.

14. *Thelypteris paleacea* A. R. Sm.
　　Sec. for., 1400–1700 m: M 5217; P 3597*.

15. *Thelypteris rudis* (Kunze) Proctor
　　Sec. for., 1200–1900 m: 27028*, 27764*; W 2614.

16. *Thelypteris scalaris* (H. Christ) Alston
　　Sec. for., 1200 m; fronds to 2 m long: V 12293*.

VITTARIACEAE (2) (* det. A. R. Smith)
Ref.: Tryon, R. M., & R. G. Stolze. 1989. Fieldiana Bot. II. 22 (Pteridophyta of Peru II, 13): 83–93.

1. *Polytaenium lineatum* (Sw.) J. Sm.
Sec. for., 1200–2100 m; ep.: 27504, 27643*; P 3584; W 2603*, 2641, 2655, 2692. [*Antrophyum lineatum* (Sw.) Kaulf. in CVPE.]

2. *Radiovittaria gardneriana* (Fée) E. H. Crane
Prim. & sec. for., 1350–2750 m; common ep.: 28017*, 28213*, 28873, 28894*, 29481, 30142*, 30301*, 30403*; P 3585; W 2654. [*Vittaria gardneriana* Fée in CVPE.]

3. *Radiovittaria remota* (Fée) E. H. Crane
Sec. for., 1250–1400 m; ep.: 29326*, 32350; W 2640, 2676, 2695. [*Vittaria remota* Fée in CVPE.]

ANGIOSPERMAE: DICOTS

ACANTHACEAE (4) (* det. L. McDade)
Ref.: Leonard, E. C. 1951–58. Contr. U. S. Nat. Herb. 31: 1–781.

1. *Blechum pyramidatum* (Lam.) Urb.
Sec. for., 1300–1500 m; weed, fls. white: 27430*, 27870; Z 173. "Barrejorno"

2. *Hygrophila costata* Nees
Sec. for., clearings, 1200–1400 m; herb, fls. white: 31105; Z 175, 205.

3. *Justicia pectoralis* Jacq.
Sec. for., 1350 m; herb, fls. white with purplish spots: Z 172.

4. *Justicia secunda* Vahl
Sec. for., 1300–1500 m; scrambler with pink fls.: 27363, 27519, 28307*; Z 123.

5. *Justicia* sp. ?
Sec. for., 1200–1300 m; herb 1 m, fls. white: 27796.

6. *Mendoncia* cf. *orbicularis* Turrill
   Prim. & sec. for., 1250–1800 m; vine with white fls.: 27597, 27683, 27737, 27762; Z 143.

ACTINIDIACEAE (1)
   Ref.: Soejarto, D. 1982. Fl. Ecuador 17: 1–48.

1. *Saurauia herthae* Sleumer
   Prim. for., 1850–1900 m; tree 5 m: F 1010.

2. *Saurauia lehmannii* Hieron.
   Prim. & sec. for., Cerro Negro, 1925–2025 m; shrub 3–4 m with white fls.: 29209, 30461, 31189. [32826 from Auca, at 2250 m, may tentatively be placed here.]

3. *Saurauia* cf. *prainiana* Buscal.
   Sec. for., 1200–1700 m; common tree 6–8 m, fls. white: 27001, 27779, 32762. "Huevo frito", "Moco"

4. *Saurauia tomentosa* (Kunth) Spreng.
   Sec. for., 1900–2450 m; shrub or tree 2–10 m, fls. white: F 1042; Sparre 14861; Zak & Jaramillo 3211, 3217, 3242. [Specimens have been determined as both var. *sprucei* (Sprague) Soejarto & var. *tomentosa*; perhaps conspecific with specimens determined as *S. prainiana*.]

AMARANTHACEAE (5)
   Ref.: Eliasson, U. 1987. Fl. Ecuador 28: 1–137.

1. *Alternanthera mexicana* (Schltdl.) Hieron.
   Weedy areas, 1200–1400 m: 27195, 28111, 30335; Filskov 37030. "Escancel"

2. *Alternanthera porrigens* (Jacq.) Kuntze
   Sec. for. & scrub, 1200–2500 m; scrambler, fls. pinkish: 27601, 28090, 28801, 32466; Z 230. [27601 & 28801 represent var. *porrigens*; 28090 & 32446 represent var. *piurensis* (Standl.) Eliasson.] "Moradilla"

3. Amaranthus hybridus L.
   Open areas, Nanegalito, 1400 m: Filskov 37016. "Ataco"

4. Amaranthus viridis L. ?
   Recent clearing, Maquipucuna, 1400 m: Z 184.

5. *Chamissoa altissima* (Jacq.) Kunth

Sec. for., 1300–2000 m; shrub or scrambler, fls. white or reddish: 27146, 27237, 27568, 30309, 31614, 31670; F 1111; G 73229; T 201, 202.

6. *Cyathula achyranthoides* (Kunth) Moq.

Sec. for. & scrub, 1200–1500 m; common weed to 1.5 m: 27207, 27409, 28290; Z 68; Filskov 37122 (det. T. M. Pedersen). "San Gémula"

7. *Iresine diffusa* Humb. & Bonpl. ex Willd.

Sec. for. & scrub, extending into prim. for., 1200–2300 m; scrambler or vine to 15 m: 27063, 27736, 28139, 28286, 28757, 28759, 30214, 32447; Z 215. "Camarón", "Escancel"

ANNONACEAE (2) (* det. L. Westra; ** det. P. J. M. Maas et al.)

Refs.: Maas, P. J. M., E. A. Mennega, & L. Y. T. Westra. 1987. Annonaceae Newsletter 4: 1–135. Westra, L. Y. T. 1995. Bot. Jahrb. 117: 273–297. Chatrou, L. W., P. J. M. Maas, & C. P. Repetur. p. 97–122 *in* R. Valencia & H. Balslev, Estudios sobre Diversidad y Ecología de Plantas. 1997.

1. *Guatteria* cf. *tessmannii* R. E. Fr.

Sec. for., 1350–1415 m; tree 20–30 m, fls. greyish brown with yellow center, frs. purplish: 31042, 31318.

2. *Raimondia cherimolioides* (Triana & Planch.) R. E. Fr.

Prim. & sec. for., 1200–1800 m; common tree 5–20 m, fls. yellow: 27060*, 27917, 27932**, 28680**, 30000**, 30442**, 31115, 31159, 31326, 31554, 31691; C 13117*; Z 165. "Anonilla"

APIACEAE (6)

Ref.: Mathias, M. E., & L. Constance. 1976. Fl. Ecuador 5: 1–71.

1. *Arracacia* cf. *elata* H. Wolff

Disturbed scrub, Quebrada Chiquilpe, 2300 m; fls. yellow-green: 32450 [atypical in leaf segments narrowly incised, inflors. entirely glabrous].

[Daucus montanus Spreng. has been collected at 01' S, 2750–2800 m, Øllgaard 98985, and may be expected in our area.]

2. *Eryngium foetidum* L.

Weed, sec. for., 1500–1600 m: 27814. "Culantrillo del monte"

[*Hydrocotyle bonplandi* A. Rich. has been collected at 78° 34'W, 2500 m: L 13657 and may be expected in our area.]

3. *Hydrocotyle humboldtii* A. Rich.
Upper montane for., 2650–2750 m; scrambler, fls. green: 29486, 30552, 31520. "Orejuela"

4. *Hydrocotyle leucocephala* Cham. & Schltdl.
Banks in sec. for., 1300–1350 m; creeping herb, fls. greenish: 27432, 28183; Z 30. "Paraguita"

5. *Hydrocotyle pusilla* A. Rich.
Sec. for., 1300–1500 m; mat-forming, frs. orange: 30319, 31750.

6. *Neonelsonia acuminata* (Benth.) J. M. Coult. & Rose ex Drude
Sec. upper mont. for., 2700–2750 m; scrambling herb to 1.5 m, fls. greenish-yellow: K 31993.

7. *Sanicula liberta* Cham. & Schltdl.
Sec. for., 1200–1500 m; fls. greenish: 27205, 27650; C 12407; K 31797; Filskov 37123 (det. B. Øllgaard).

8. *Spananthe paniculata* Jacq.
Pastures, 1375–1400 m: Z 207, 224.

APOCYNACEAE (3)
Ref.: Woodson, R. E. 1933. Ann. Missouri Bot. Gard. 20: 605–790; 1936. ibid. 23: 153–206.

1. *Mandevilla callista* Woodson
Sec. for., 1200–1400 m; vine with white latex; corolla white with purple-rimmed lobes: 31098, 31319.

2. *Mandevilla veraguasensis* (Seem.) Hemsl.
Sec. for., Sendero de las Palmitas, 1300 m; dull purplish fls. on trail: 32739.

3. *Mesechites trifida* (Jacq.) Müll. Arg.
Sec. for., 1130–1250 m; vine with greenish-white or yellowish-white fls.: 31270, 31723; V 12270 (det. B. Hansen); Z 19, 198.

4. *Prestonia* cf. *trifida* (Poepp.) Woodson ex Gleason & A. C. Sm.
Sec. for., west of Marianitas, 1150–1300 m; rare vine, fls. purplish: 28834.

AQUIFOLIACEAE (1)
Ref.: Edwin, G. 1965. Mem. N. Y. Bot. Gard. 12(3): 124–150.

1. *Ilex* sp. (aff. *I. nervosa* Triana ?)
Sec. for., 1550 m; treelet: G 73127.

ARALIACEAE (3) (* det. F. Borchsenius)
Ref.: Borchsenius, F. 1997. Nordic J. Bot. 17: 373–396. Ulloa U., C., & P. M. Jørgensen. 1993. Árboles y arbustos de los Andes del Ecuador: 64–66.

1. *Dendropanax* cf. *macrophyllum* Cuatrec.
Sec. for., 1200–1950 m; tree 13–20 m high, fls. greenish: 27626, 28258, 32970; V 12310. "Malva"

2. *Dendropanax* sp. 1
Sec. for., 1550–1630 m; tree: G 69935, 73166.

3. *Oreopanax confusus* Marchal
Prim. & sec. for., 1650–2100 m; liana or tree to 16 m: 29338*, 30257, 31775; 32772, C 5929*; F 1080, 1107, 1205, 1216, 1512; G 73210*; P 3602*; T 180*.

4. *Oreopanax* cf. *eriocephalus* Harms
Prim. & sec. for., 1300–1700 m; tree 10–20 m: 31708, 32950.

5. *Oreopanax floribundus* (Kunth) Decne. & Planch.
Loma Pahuamba, 1750 m; tree 5–15 m: F 1266, 1582. [28395, a sterile specimen from a sapling, may belong here.]

6. *Oreopanax grandifolius* F. Borchsenius
Prim. & sec. for., 1200–1700 m; tree 5–10 m: C 39762; G 69934*; Q 54*; T 142*. "Pumamaqui"

[*Oreopanax palamophyllus* Harms has been collected 02′ south of our limits, 2500 m, P 3626*.]

7. *Oreopanax* sp. 1
Upper mont. for., Cerro Montecristi, 2725 m: 29499.

8. *Schefflera dielsii* Harms
Sec. for. to prim./sec. for. transition, 1400–1650 m; vine-like shrub or small tree to 5 m: 31134, 32951; Z 85.

9. *Schefflera* cf. *ferruginea* (Willd. ex Roem. & Schult.) Harms
Prim. & sec. for., 1200–1750 m; common treelet 2–3 m: 27170, 27642, 28041, 28729, 28853, 29243, 30420; F 1503, 1598; G 69944, 73162.

10. *Schefflera lasiogyne* Harms
Prim. & sec. for., 1200–2000 m; shrub 4 m, fls. greenish-yellow: 31075; C 5915 (det. D. Frodin), 13083*; F 1214; Q 17*; T 169*, 203*; V 12253*, 12306*. [28956, a tree 15 m tall at 2050–2100 m, may belong here.]

11. *Schefflera* cf. *sphaerocoma* (Benth.) Harms
Sec. for., 1600 m; tree 8 m, inflors. c. 0.75 m, fls. whitish-yellow: 31765.

12. *Schefflera* cf. *whitefoordiae* M. J. Cannon & Cannon
Sec. for., 1200–1250 m; tree to 7 m: 28972; V 12294. [Perhaps not distinct from *S. sphaerocoma.*]

ASCLEPIADACEAE (6)
Ref.: Spellman, D. L. 1975. Ann. Missouri Bot. Gard. 62: 103–156.

1. *Asclepias curassavica* L.
Common weed, sec. for., 1200–1300 m; prim. for., 1900 m: 27307, 27658, 30127. "Viborana"

2. *Cynanchum* sp.
Prim. for., 2100 m; small vine, fls. yellow: P 3594.

3. *Fischeria columbiana* Schltr.
Sec. for., betw. Marianitas & Nanegal, 1200–1250 m; vine, fls. greenish-white: 28312. [Not in CVPE.]

4. *Gonolobus* sp. ? [genus doubtful]
Prim. for., 1925 m; vine, fls. white: 29222.

5. *Matelea* cf. *ecuadorensis* (Schltr.) Morillo
Sec. for., 1300–1700 m.; vine, fls. green; frs. green, woody, 20 cm long: 27151, 31333. [According to CVPE, known only from the type.]

6. *Matelea lehmannii* (Schltr.) Morillo
   Sec. for., 1220 m: Croat 38890 (det. G. Morillo).

7. *Oxypetalum cordifolium* (Vent.) Schltr.
   Swampy area betw. El Carmen & Marianitas, 1150 m; vine in thickets, not
   common, fls. yellow: 29076.

ASTERACEAE (53) (* det. H. Robinson; ** det. J. Strother)
   Refs.: Aristeguieta, L. 1964. Fl. Venezuela 10: 1–941. D'Arcy, W. G. (ed.).
   1975. Ann. Missouri Bot. Gard. 62: 835–1322. Ulloa U., C., & P. M.
   Jørgensen. 1993. AAU Rep. 30: 70–102. Robinson, H., & V. A. Funk. 1997.
   Compositae of Ecuador, I: key to frequently collected genera. Mem. II
   Congreso Ecuatoriano Bot. 65–78.

1. *Acmella alba* (L'Hér.) R. K. Jansen var. *ecuadorensis* R. K. Jansen
   Weed in sec. for., fls. white, 1150–1800 m: 27415**, 27886**, 28826**,
   32787; C 13023; Z 195; Filskov 37057. "Batoncillo"

2. *Adenostemma platyphyllum* Cass.
   Sec. for., 1200–1500 m; weed with white fls.: 27377; Filskov 37114*. [Some
   Ecuador specimens have been determined as *A. lavenia* (L.) Kuntze, a species
   not confirmed for the neotropics.] "Tía Juana", "Mama Juana"

3. *Ageratina psilodora* (B. L. Rob.) R. M. King & H. Rob. ?
   Sec. for., 1900–1925 m; herbaceous weed, fls. white: 31619*.

4. *Ageratum conyzoides* L.
   Sec. for., 1300–1500 m: 27404; Z 50. "Pedrorera"

5. *Ambrosia arborescens* Mill.
   Sec. for., 1400 m: Filskov 37043.

6. *Ayapanopsis cuchabensis* (B. L. Rob.) R. M. King & H. Rob.
   Sec. for., 1750–2250 m; shrubby herb to 2 m, corolla lavender to dark blue:
   31933*, 32841, 32861.

7. *Baccharis jelskii* Hieron.
   Sec. for., 1250–1700 m; shrub or vine, fls. white, fragrant: 27006, 29251*,
   29342*, 31695.

8. *Baccharis latifolia* (Ruiz & Pav.) Pers.
   Upper mont. for., 2250–2750 m; shrub 2–3 m, fls. whitish: 32451; H 29539*; L 13659. "Chilca"

9. *Baccharis nitida* (Ruiz & Pav.) Pers.
   Sec. for., 1200–1250 m; shrub or tree to 6 m, fls. white: 27313; Z 46; Ward s.n.

10. *Baccharis pululahuensis* Hieron.
    Sec. for., 1250–1750 m; scandent shrub, fls. white: 28332*, 30412*, 31652*, 31693, 31935*.

11. *Baccharis teindalensis* Kunth
    Prim. for., 1800–2000 m; shrub 4 m: C 7190*.

12. *Baccharis trinervis* Pers.
    Rip. for., 1350–1800 m; shrub 1.5–3 m, fls. white: 27587; C 13040; Z 2. [Maquipucuna specimens are atypical in their lanceolate rather than elliptic leaves.] "Chilca"

13. *Barnadesia parviflora* Spruce ex Benth. & Hook. f.
    Sec. for., 1100–1900 m; spiny shrub 3–5 m, fls. pink: 29269*, 30228, 30514*; F 1520; Holm-Nielsen 24428 (det. E. Urtubey). "Espino de gato"

14. *Bidens pilosa* L.
    Sec. for., 1200–2400 m; common herbaceous weed: 27238, 27431, 30205; Z 220; Filskov 37067. "Crespa morada", "Moriseco"

15. *Bidens squarrosa* Kunth
    Prim. & sec. for., 1500–1700 m; climbing or creeping vine: 28033, 31759**, 32948.

16. *Bidens triplinervia* Kunth
    Disturbed areas, 2000 m; herb, fls. yellow: 31864**.

17. *Centratherum punctatum* Cass.
    Sec. for., betw. Marianitas & Nanegal, 1150 m; weed, fls. blue: 28821.

18. *Chaptalia nutans* (L.) Polak
    Sec. for., 1250–1800 m; acaulescent herb, bracts purplish-tipped, fls. pink: 27384, 27426; C 14232*; Z 190.

19. *Chromolaena scabra* (L. f.) R. M. King & H. Rob.
Sec. for., 1700–1925 m; shrub 1.5–2 m, fls. pale lavender: 31583, 31936*, 32768.

20. **Clibadium alatum** H. Rob.
Sec. for., 1550 m; tree 6 m: 29256*.

21. *Clibadium eggersii* Hieron.
Sec. for., Cerro Santa Lucia, 1650 m; shrub 1.5 m, fls. white: 30352*.

22. *Clibadium laxum* S. F. Blake
Sec. for., 1250–2000 m; shrub or tree to 5 m, fls. whitish, frs. yellow & fleshy: 27314*, 28745*, 30608*, 31718*; G 73205*; N 9802*; Z 229.

23. *Clibadium surinamense* L.
Sec. for., 1500–2750 m; shrub, often scrambling, 1.5–2.5 m, fls. white: 30483*, 31951*, 32807, 32838; K 31967*.

24. *Clibadium sylvestre* (Aubl.) Baill.
Sec. for., Cerro Santa Lucia, 1500 m; shrub 3 m, fls. white: 28327*.

25. *Clibadium websteri* H. Rob.
Sec. for., Cerro Negro, 1750–2050 m; shrub with cane-like stems to 3 m: 30474*, 31930* [type collection]; [aff. *Clibadium trianae* (Hieron.) S. F. Blake, fide H. Robinson (1997)].

26. *Conyza bonariensis* (L.) Cronquist
Clearings in sec. for., 1500–1600 m; herbaceous weed: 32117.

27. *Critonia morifolia* (Miller) R. M. King & H. Rob.
Sec. for. & scrub, 1300–1500 m; scrambling to 4 m, fls. whitish, fragrant: 31696*, 32356.

28. *Critoniopsis occidentalis* (Cuatrec.) H. Rob.
Prim. & sec. for., 1350–2000 m; tree 10–20 m, fls. purplish: 28689*, 30246*, 31314; Q 7*, 31*, 285*.

29. *Dendrophorbium lloense* (Hieron.) Jeffrey [*Senecio lloensis* Hieron.]
Prim. & sec. for. to upper mont. for., 1400–2750 m; shrub 3–4 m, rays yellow: 30453, 30601*, 31950*, 32766; H 29548*.

30. *Elephantopus mollis* Kunth
    Weed, fls. whitish, 1200–1400 m: 27087; Z 15; Filskov 37065 (det. B. Øllgaard).

31. *Eleutheranthera ruderalis* (Sw.) Sch. Bip.
    Weed in sec. for., 1250 m; fls. yellow: Hunter s.n.

32. *Erato polymnioides* DC.
    Sec. for. & scrub, 1200–1700 m; coarse weedy herb 1–2 m, fls. yellow: 27068, 27309, 27586, 28177, 32770; C 12416.

33. *Erechtites hieraciifolius* (L.) Raf. ex DC.
    Common weed to 1 m, fls. yellow, 1400–1500 m: 27183, 27199; Z 186.

34. *Fleischmannia huigrensis* (B. L. Rob.) R. M. King & H. Rob. ?
    Sec. for., 22 km from Nono towards Nanegalito, 1850 m; liana, with pale violet rays: Grimes & Todzia 2488.

35. *Fleischmannia obscurifolia* (Hieron.) R. M. King & H. Rob.
    Weed, fls. lavender, 1200–1900 m: 27086, 30122*; Z 51; Filskov 37059.

36. *Galinsoga quadriradiata* Ruiz & Pav.
    Sec. for., 1150–2800 m; weed, rays pink or white, disk yellow: 27433, 28829*, 30108, 31507, 32785.

37. Gnaphalium luteo-album L. ?
    Sec. for., weed, 1500–1600 m: 27567, 32777.   [Determination doubtful.] "Hierba blanca", "Lechugilla"

38. *Gnaphalium purpureum* L. [*Gamochaeta purpurea* (L.) Cabrera]
    Sec. for., weed, 1900 m: 30119.

39. *Hebeclinium killipii* (B. L. Rob.) R. M. King & H. Rob.
    Prim. & sec. for., 1300–1875 m; shrub to 4 m, sometimes clambering, fls. whitish or purple: 27484*, 29084*, 29240*, 29365, 30341*.

40. *Hebeclinium obtusisquamosum* (Hieron.) R. M. King & H. Rob.
    Prim. for., 1800 m; tree 6 m, fls. white: 28762*.

41. *Hebeclinium tetragonum* Benth.
    Scrub on banks, 2000–2350 m; shrub to 2.5 m, fls. blue: 31539; K 32020.

42. *Heliopsis buphthalmoides* (Jacq.) Dunal
Sec. for., 1150–1700 m; herb with yellow fls.: 27378**, 27434**, 28118*, 28804*, 28831**, 32759; Filskov 37052*. [A number of collections have been determined as *H. oppositifolia* (Lam.) Diaz, an illegitimate name.]

43. *Heterocondylus vitalbae* (DC.) R. M. King & H. Rob.
Sec. for., 1200–1400 m; somewhat scandent to 3 m, fls. pinkish or lilac: 27069, 28343*, 32909; D 6080; Z 23.

[*Hieracium frigidum* Wedd., between Calacalí & Nanegalito, 2175 m, Croat 72876*, is just east of our boundary.]

44. *Hypochaeris chillensis* (Kunth) Hieron.
Roadsides, 2000–2300 m; fls. yellow: 31812, 32819.

45. Hypochaeris radicata L.
Trailsides, 2250 m; fls. yellow: 32831.

46. *Jaegeria hirta* (Lag.) Less.
Common weed, 1200–2750 m; fls. yellow: 27439, 30539*, 32786; Filskov 37079*.

47. *Jungia mitis* Benoist
Sec. for., 1950–2000 m; climbing vine, fls. white, fragrant: 32444.

48. *Liabum stipulatum* Rusby
Scrub on banks, 1500–2000 m; herb 0.5–1 m, fls. yellow: 31856, 31914, 32783.

49. *Mikania banisteriae* DC.
Sec. for., 1200–1550 m; vine to 5 m, fls. white: 27005 (det. R. Noyes), 27639 (det. A. J. Bornstein); G 73131*.

50. *Mikania congesta* DC.
Sec. for., 1920 m; creeping vine, fls. white: 30125*.

51. *Mikania cordifolia* (L. f.) Willd.
Sec. for., 1250–1300 m; common weedy vine, fls. white: 27016; N 9792*; Z 18. "Guaco"

52. *Mikania leiostachya* Benth.
    Prim. & sec. for., 1300–2000 m; vine with whitish fls.: 27286, 29340*, 30247, 30419, 31932.

53. *Mikania lloensis* Hieron.
    Prim. for., 2100–2150 m; vine with pink fls.: 28885*.

54. *Mikania micrantha* Kunth
    Sec. for., 1650–2200 m; liana, fls. white: 32388, 32776; Croat 72893*; Grimes 2490 (det. B. L. Turner); Zak & Jaramillo 3226*.

55. *Mikania sylvatica* Klatt
    Prim. for., Cerro Negro, 1950–2000 m; clambering vine, fls. yellowish-white: 30470*.

56. *Munnozia hastifolia* (Poepp.) H. Rob. & Bretell
    Sec. for., 1200–1500 m; coarse herb to 1.5 m, fls. yellow: 27083, 29061, 31725, 31912.

57. *Munnozia liaboides* (Less.) H. Rob.
    Sec. for., 1250 m; coarse herb, fls. yellow: 31713* [specimen collected with *M. pinnatipartita*, appearing intermediate between *M. hastifolia* & *M. pinnatipartita*].

58. *Munnozia pinnatipartita* (Hieron.) H. Rob. & Bretell
    Sec. for., 1250–2000 m; coarse herb, fls. yellow: 27056*, 31714, 32425, 32800. [31726 appears intermediate between *M. hastifolia* & *M. pinnatipartita*.]

59. *Munnozia senecionidis* Benth.
    Prim. & sec. for., 1600–2750 m; vine to 15 m, fls. yellow: 27832, 28704, 29461, 30459, 30490, 31126, 32840; H 29540*.

60. *Mutisia grandiflora* Bonpl.
    Banks, 2350 m; vine with pendulous red fls.: K 32015.

61. *Neomirandea homogama* (Hieron.) H. Rob. & Bretell
    Sec. for., 1600–1725 m; herb 1–2 m, fls. purplish or pinkish: 28349*, 31685*.

62. *Oligactis pichinchensis* (Hieron.) H. Rob. & Bretell
    Prim. & upper mont. for., 2000–2750 m; vine with yellow fls.: 28944*, 30185; H 29529*.

63. *Pentacalia floribunda* Cuatrec.
    Sec. scrub, 2350 m; shrub to 2.5 m, lvs. purplish beneath, phyllaries purplish, fls. yellow: K 32016*.

64. *Pentacalia huilensis* (Cuatrec.) Cuatrec.
    Prim. & sec. for., 1250–2275 m; woody vine, lvs. often purplish beneath, fls. white: 29152, 29339, 29440*, 32903, 32926. [G 73186 may also belong here.]

65. *Pentacalia jelskii* (Hieron.) Cuatrec.
    Prim. & sec. for., 1250–2250 m; vine with yellowish-white fls.: 27739*, 28943*. [Not in CVPE.]

66. *Pentacalia luteynorum* H. Rob. & Cuatrec. ?
    Prim. & upper mont. for., 2250–2750 m; climber with yellow fls.: 29439*; H 29532*.

67. *Pentacalia riotintis* (Cuatrec.) Cuatrec. ?
    Upper mont. for., 2700–2750 m; shrub to 3 m, fls. yellow: K 31973*. [Previously known only from the type, according to CVPE.]

68. *Piptocoma* cf. *discolor* (Kunth) Pruski
    Sec. for., 1650 m; tree 4 m: 29262 [specimen barren, species identity needing confirmation].

69. *Polyanthina nemorosa* (Klatt) R. M. King & H. Rob.
    Sec. for., 1150–2100 m; weedy herb 1–2 m, fls. white or pale blue: 27091, 27311*, 28833, 29196*, 29354, 30468*, 31332, 32967; Z 214.

70. *Pseudelephantopus spicatus* (Juss. ex Aubl.) J. B. Rohr
    Sec. for., 1300–1350 m; weed with pinkish fls.: 27417; Z 35. "Chicoria"

71. *Pseudelephantopus spiralis* (Less.) Cronquist
    Sec. for., 1200–1400 m; weed with pinkish fls.: 29068; Filskov 37078*.

72. *Pseudogynoxys sonchoides* (Kunth) Cuatrec.
    Sec. for., 1250 m; decumbent, fls. orange: 29033*.

73. *Schistocarpha eupatorioides* (Fenzl) Kuntze
    Sec. for., 1200–1250 m; herb to 1.5 m, fls. yellow: 27085.

74. *Sigesbeckia jorullensis* Kunth
    Sec. for., clearings, 1275–1800 m; herb with yellow fls.: 30221, 31883; Z 206;
    Filskov 37126*.

75. *Smallanthus riparius* (Kunth) H. Rob.
    Sec. for., 1150–1250 m; herb to 2 m, fls. yellow: 28297*, 28802*. "Polaco"

    [Sonchus oleraceus L., from roadsides between Calacalí & Nanegalito, 2250
    m, Croat 72859, is 2' east of our boundary.]

76. *Stevia* cf. *andina* B. L. Rob.
    Prim. for., Auca, 2250 m; herb 0.5 m, fls. white: 32824.

77. *Stevia elatior* Kunth
    Mossy banks, 2000–2250 m; herb 0.3–0.5 m, fls. white: 31857.

78. *Tagetes multiflora* Kunth
    Banks, 2300 m; annual herb, fls. yellow: K 31962. "Tzintso"

79. *Tagetes terniflora* Kunth
    Sec. for., 1400 m: Filskov 37038*. "Asnayuyo"

80. Taraxacum officinale Weber ex F. H. Wigg.
    Roadsides, 2000 m; fls. yellow: 31863. "Diente de león"

81. *Tessaria integrifolia* Ruiz & Pav.
    Sec. for., 1200–1250 m; common shrub c. 3 m, fls. white: 28293, 28973,
    31090.

82. *Tilesia baccata* (L.) Pruski
    Sec. for., 1150–1350 m; shrub or scrambler, fls. yellow: 27065, 28815.
    [Previously known as *Wulffia baccata* L.; transferred by Pruski in Novon 6:
    414. 1996.]

83. *Verbesina arborea* Kunth
    Prim. for., 1200–1700 m; tree 15–20 m × 30 cm: 31780; T 145*.

84. *Verbesina nudipes* Blake
Sec. scrub & woodland, Quebrada Chiquilpe, 2200–2350 m; bushy shrub or tree 3–4 m, fls. white: 32452, 32453; K 32018. [Not in CVPE.]

85. *Vernonanthura patens* (Kunth) H. Rob.
Sec. for., 1200–1250 m; common shrub c. 3 m, fls. white: 27072; Z 1, 91.

86. *Vernonia* sp.
Sec. for., 1550 m; tree: G 73190.

87. *Viguiera quitensis* (Benth.) S. F. Blake
Lower & upper mont. scrub, 2000–2350 m; shrub to 1.5 m, fls. yellow: 31861*; K 32017*.

BALANOPHORACEAE (3)
Ref.: Hansen, B. 1983. Fl. Ecuador 19: 1–16.

1. *Helosis cayennensis* (Sw.) Spreng.
Sec. for., 1175–1350 m; plants fleshy, pink to red: 27389, 27929, 27969, 28812, 32383.

2. *Langsdorffia hypogaea* Mart.
Prim. for., 1900–2000 m; involucre reddish-brown: 31639, 31746; F 1195.

3. *Scybalium depressum* (Hook. f.) Eichler
Prim. for., 2250–2500 m; stems & scales red, fls. yellow: 29153, 29457, 30586.

BALSAMINACEAE (1)

1. Impatiens walleriana Hook. f.
Spreading on roadsides & trails, 1200–1300 m: 28355. "Mirame linda"

BEGONIACEAE (1) (* det. L. B. Smith; ** det. S. F. Smith)
Refs.: Smith, L. B., & D. C. Wasshausen. 1986. Fl. Ecuador 25: 1–65. Smith, L. B., et al. 1986. Smiths. Contr. Bot. 60: 1–584.

1. *Begonia bracteosa* A. DC. ?
Sec. for., 1800 m; vine: C 13006 [not in CVPE; determination doubtful].

2. *Begonia exalata* C. DC.
   Prim. for., 21 km west of Calacalí, 2500 m: L 13670*.

3. *Begonia foliosa* Kunth
   Prim. & sec. for., esp. trails & banks, 1200–2000 m: 27254, 27592, 27786, 27912, 28278, 29310, 30194, 30424; Flynn 4123; Hurtado 1417 (det. M. Bjerrum). "Llorona"

4. *Begonia* cf. *geminiflora* L. B. Sm. & Wassh.
   Prim. & sec. for., 1450–1500 m; ep. vine: 31894. [Cited only from the type in CVPE; agrees with *B. dodsonii* in some respects, but the large-winged fruit & long pedicel suggests greater resemblance with *B. geminiflora*.]

5. *Begonia glabra* Aubl.
   Sec. for., 1300–1400 m; fls. white: 29289, 29335. "Hoja de sapo"

6. *Begonia longirostris* Benth.
   Prim., sec. & upper mont. for., often gregarious on wet banks, 1300–2500 m; terr., fls. red: 27259, 27282, 27735, 28044, 28935, 29145, 29245, 30385, 30460, 30572, 31594, 31830, 32810; F 1028; Harling & Andersson 11608; Holm-Nielsen 24449*.

7. *Begonia maurandiae* A. DC. ?
   Sec. for., 1750 m; ep. vine: 30057**.

8. *Begonia parviflora* Poepp. & Endl.
   Sec. for., 1250–1900 m; common, treelike, 2–5 m high, fls. white: 27003, 27465, 28259, 28260, 29047, 30358, 31948; C 39725; G 73179; N 9780; Z 100.

9. *Begonia pululahuana* C. DC.
   Prim. for., 1900–2000 m; ep., fls. pink & yellow: 30225**, 32406.

10. *Begonia secunda* L. B. Sm. & Wassh.
    Prim. for., 1750–1900 m; ep. vine, fls. red: 27326, 28850, 31624. [Cited only from the type in CVPE; difficult to distinguish from *B. maurandiae*.]

11. *Begonia sodiroi* C. DC.
    Prim., sec., & upper mont. for., 1300–2750 m; ep. vine, fls. pink: 29104, 31525, 31841.

12. *Begonia tiliifolia* C. DC.
Prim. for., 1525–2050 m; fls. white: 27138*, 28218, 28690, 29200, 30149, 31340, 32437.

13. *Begonia truncicola* Sodiro ex C. DC.
Prim. for., 2100–2275 m; vine, fls. red: 28886, 29161**, 29446**, 30045.

14. *Begonia* sp. 1
Prim. & sec. for., 1500–2175 m; ep. vine, fls. pink: 27957.

15. *Begonia* sp. 2
Sec. for., 1550 m; terr., fls. white: G 73248.

BETULACEAE (1)
Ref.: Furlow, J. J. 1979. Rhodora 81: 1–121, 151–248.

1. *Alnus acuminata* Kunth ssp. *acuminata*
Woods along streams, collected at 1200–1300 m but observed above 2000 m; tree to 10 m: 30446; C 6863. "Aliso"

BIGNONIACEAE (6) (* det. A. Gentry)
Ref.: Gentry, A. H. 1977. Fl. Ecuador 7: 1–173.

1. *Arrabidaea chica* (Bonpl.) B. Verl.
Sec. for., 1300–1350 m; liana in canopy, fls. lilac: 31034; N 9781*. "Corivasca"

2. *Delostoma integrifolium* D. Don
Sec. for., 1350–2200 m; tree 5–20 m, fls. purple-violet: 29111, 31186, 31853; F 1058, 1483; Croat 72864*; Rubio 2417. "Yaloman"

3. *Pithecoctenium crucigerum* (L.) A. H. Gentry
Sec. for., 1350 m; liana, fls. yellow: 31082. "Cucharilla"

4. *Schlegelia monachinoi* Moldenke
Sec. for., El Carmen to Marianitas, 1200–1250 m; tree 6 m, corolla white with purplish limb: 27300*.

5. *Schlegelia sulphurea* Diels ?
Sec. for., liana, fls. white, 1300–1400 m: 27518.

6. *Tourrettia lappacea* (L'Hér.) Willd.
   Rip. for., 1900–2000 m; climber, fls. red: 30227, 31590, 32435.

7. *Tynanthus schumannianus* (Kuntze) A. H. Gentry
   Sec. for., 1630 m; tree, fls. white: G 69902*. [Not in CVPE, cf. *T. polyanthus*
   (Bureau) Smith.]

BIXACEAE (1)

1. Bixa orellana L.
   Cult. shrub or tree, occ. in sec. for., c. 1200–1400 m: 31174; Filskov 37008.
   "Achiote"

BOMBACACEAE (5) (* det. W. Alverson)
   Ref.: Alverson, W. S. 1989. Taxon 38: 377–388.

1. *Ceiba* aff. *salmonea* (Ulbr.) Bakh.
   Sec. for., 1200–1550 m; tree to 30 m, trunk spiny: 29090, 31165, 31265. [Not
   in CVPE.]

2. *Matisia castano* H. Karst. & Triana
   Sec. for., 1500–1700 m; tree 7 m: Q 117*.

3. *Matisia* cf. *malacocalyx* (Robyns & S. Nilsson) Alverson
   Prim. & sec. for., 1550–1800 m; tree 8–15 m, fls. whitish: 28211, 29372,
   31146; C 39751; F 1521, 1529; G 69939, 73122, 73193; Q 116; T 194.

4. *Matisia* cf. *soegengii* Cuatrec.
   Rip. for. along Río Umachaca, 1250 m; tree 25 m: 31711 [specimen barren,
   requiring verification].

5. *Matisia* sp. 1
   Sec. for., Hacienda El Carmen, 1250 m; tree 18 m × 20 cm, fls. white: 31645.

6. *Ochroma pyramidale* (Cav. ex Lam.) Urb.
   Sec. for., 1500–1600 m; tree 20–25 m: 31654, 32960. "Balsa"

7. *Pachira patinoi* (Dugand & Robyns) Fern. Alonso
   Sec. for., 1500–1700 m; tree 12 m: G 73132; Q 101.

8. *Spirotheca* sp.
Sec. for., 1200–1700 m; tree, spiny trunk: G 73183*; T 197*.

BORAGINACEAE (4) (* det. J. S. Miller)
Ref.: Miller, J. S. 1988. Ann. Missouri Bot. Gard. 75: 456–521; 2000. Novon 10:48–52.

1. *Cordia cylindrostachya* (Ruiz & Pav.) Roem. & Schult.
Sec. for., 1250–2400 m; shrub 2–7 m, fls. white: 32803; N 9785*; Q 335*; Zak & Jaramillo 3236*.

2. *Cordia cymosa* (J. D. Sm.) Standl.
Sec. for., 1300–1550 m; tree 7 m, frs. greenish-grey: 31022. [G 73160 probably belongs here.]

3. *Cordia hebeclada* I. M. Johnst.
Sec. for., 1300 m; tree 20 m: N 9787.

4. *Cordia mexiana* I. M. Johnst.
Sec. for., 1500–1700 m; tree 10 m, frs. red: Q 296*.

5. *Cordia* aff. *spinescens* L.
Sec. for., 1250–1950 m; shrub or tree 3–6 m, fls. whitish or yellowish-green: 28105, 28180, 31617, 31949, 32764; Z 32, 193. [Our plants are atypical, with large entire leaves & long spikes not adnate to the petiole.]

6. Cynoglossum amabile Stapf & J. R. Drumm.
Sec. for. & clearings, 1900–2000 m; fls. blue: 28084, 30112. "No me olvides"

7. Heliotropium indicum L.
Sec. for., clearings, 1850–1900 m; herb to 1 m, fls. lavender: 31629. "Rabo de alacrán"

8. *Heliotropium rufipilum* (Benth.) I. M. Johnst.
Sec. for., 1200–1800 m; 0.5–1.5 m high, fls. white: 27074, 27637, 28146, 32745; C 12433*.

9. *Tournefortia fuliginosa* Kunth
Rip. & sec. for., 1600–2000 m; arborescent shrub 3–7 m, fls. white, with foetid odor: 31852, 31934, 32781; F 1268. "Escorpión"

10. *Tournefortia gigantifolia* Killip ex J. S. Mill.
Sec. for., 1300–1875 m; shrub with mostly unbranched stems c. 2–4 m, fls. creamy white or greenish, sometimes purplish-tinged, with foetid odor: 29100, 30359, 31667, 31758, 31945. [Not in CVPE.]

11. *Tournefortia glabra* L.
Sec. for., 1300–1400 m; shrub 1.5 m, fls. greenish: 29313, 29986, 30311.

12. *Tournefortia microcalyx* (Ruiz & Pav.) I. M. Johnst.
Sec. for., 1200–1650 m; liana, fls. white with cloying odor: 31328, 31660; T 149*.

13. *Tournefortia ovalifolia* Rusby ?
Sec. for., 1400–1425 m; liana, frs. yellow: 31266. [Not in CVPE.]

14. *Tournefortia ramosissima* K. Krause
Disturbed scrub, 1200–2300 m; shrub 2–2.5 m, fls. white: 31096, 32354.

15. *Tournefortia scabrida* Kunth
Sec. for., 1900–2200 m; shrub 3 m, fls. green, frs. white: Zak & Jaramillo 3228*.

16. *Tournefortia* sp. 1
Sec. rip. for., 1250 m; tree 15 m: 29058.

BRASSICACEAE (1)

1. *Cardamine ovata* Benth.
Rip. for., clearings, 2000–2100 m; fls. white: 32422, 32852.

BRUNELLIACEAE (1)
Ref.: Cuatrecasas, J. 1970. Fl. Neotrop. Monogr. 2: 1–189; 1985. suppl.: 28–103.

1. *Brunellia acostae* Cuatrec.
Loma Pahuamba, 2300 m; tree 7 m: F 1345.

2. *Brunellia comocladifolia* Humb. & Bonpl.
Prim. & sec. for., 1300–2000 m; tree 5–15 m: 28391, 28755, 30258; N 9789; T 143; Z 235. "Cedrillo"

BUDDLEIACEAE (1)
   Ref.: Norman, E. 1982. Fl. Ecuador 16: 1–24.

1. *Buddleia americana* L.
   Sec. for., 1250–1700 m; shrub 1–2.5 m, fls. yellow: 27002, 32742. "Salvia"

CAMPANULACEAE (4) (* det. T. Lammers)
   Ref.: Jeppesen, S. 1981. Fl. Ecuador 14: 1–170.

1. *Burmeistera crispiloba* Zahlbr.
   Prim. & sec. for., 1200–2750 m; scandent shrub 2–4 m, fls. green or purplish:
   27058, 27096, 27772, 27837, 28243, 28325, 28945, 28947, 29157, 29352,
   29443, 30017, 30455, 30545; F 1021; G 69945; N 8654; T 157. [Variable;
   some specimens with longer peduncles (e.g., 28945, 29443) may represent *B.
   sodiroana* Zahlbr.]

2. *Burmeistera glabrata* (Kunth) Benth. & Hook. ex B. D. Jacks.
   Upper mont. for., 2700–2750 m; shrub to 2 m, fls. greenish-yellow: K 31991.

3. *Burmeistera multiflora* Zahlbr.
   Prim. & sec. for., 1200–2300 m; shrub 1–3 m, fls. greenish or purplish: 27234,
   27548, 27670, 27833, 28073, 28323, 29187, 29382, 30041, 30133, 30340,
   30464, 31607, 32848; F 1017; N 8649*. [Type from Nanegal, Karsten s.n.]

4. *Burmeistera resupinata* Zahlbr. var. *heilbornii* E. Wimm.
   Prim. for., 1500–1700 m; ep. or terr., fls. green to purplish-brown: N 9800*.

   [*Burmeistera sodiro* Zahlbr., from 5 km southwest of Nieblí, Stein 2662, is
   slightly east of our boundary.]

5. *Burmeistera succulenta* H. Karst. & Triana
   Upper mont. for., 2650–2700 m; shrub 0.5–2.5 m, stems rather succulent, fls.
   greenish or purplish: 31515; K 31995.

6. *Centropogon aequitorialis* E. Wimm.
   Prim. & sec. for., 1900–2750 m; herb or subshrub, calyx green, corolla
   pinkish: 27955, 30113, 30154, 30485, 30541, 31527, 31542.

7. *Centropogon calycinus* Benth.
   Prim. & upper mont. for., 2000–2700 m; shrub to 1.5 m high, fls. magenta:
   C 5946 (det. J. Luteyn); K 32004.

[*Centropogon glabrifilis* (E. Wimm.) Jeppesen from 00°01′N, 78°34′W, 2500 m, L 13660, is just east of our boundary.]

8. *Centropogon nigricans* Zahlbr.
   Prim. & upper mont. for., 1800–2500 m; shrub 1.5–3 m, fls. greenish: 27673, 28233, 28328, 29208, 30486, 30578.

9. *Centropogon preslii* E. Wimm.
   Upper mont. for., 2250–2750 m; scarcely woody shrub to 1 m, corolla red with yellow lobes: 31526, 32835; H 29533*.

10. *Centropogon solanifolius* Benth.
    Prim. & sec. for., 1200–2175 m; shrub 1–1.5 m, fls. scarlet: 27124, 27745, 28047, 28134, 28191, 29216, 29407, 30085, 30206, 30463; F 1067; G 73253; N 8647*; T 196.

11. *Centropogon subandinus* Zahlbr.
    Sec. for. & upper mont. for., 1950–2500 m; shrub c. 2 m, fls. erect, corolla magenta; frs. deflexed: 32387; L 13634.

12. Hippobroma longiflora (L.) G. Don
    Roadsides, 1250 m; herb with white fls.: 31085. "Jazmincillo"

13. *Lobelia xalapensis* Kunth
    Sec. for., 1250–1300 m; herb with white or pale lavender fls.: 27656.

CAPPARIDACEAE (2)

1. *Cleome anomala* Kunth
   Sec. for., 2700 m; shrub 2 m, fls. yellow: K 32006.

2. *Podandrogyne* sp. (ined.)
   Prim. & sec. for., 1250–2050 m; shrub 1–3 m, fls. yellow or orange: 28330, 28691, 28707, 30118, 30217, 31028, 31184, 31559, 31570, 32958.

CAPRIFOLIACEAE (2)
   Ref.: Ulloa, C., & P. M. Jørgensen. 1993. AAU Rep. 30: 116–118.

1. *Sambucus peruviana* Kunth
   Sec. for., 1250–1500 m; shrub 2–3 m, sometimes cult., fls. white: 27473, 27715, 28773; Filskov 37006 (det. P. M. Jørgensen), 37040. "Sauco", "Tilo"

2. *Viburnum pichinchense* Benth.
Prim. & upper mont. for., 2200–2750 m; shrub or tree 5–20 m, fls. white, fragrant: 28904, 28926, 29441, 30524, 30604, 32818; H 29527; K 31997.

CARICACEAE (1)
Ref.: Badillo, V. 1983. Fl. Ecuador 20: 27–48; 2000. Ernstia 10:74–79.

1. *Carica longiflora* V. Badillo [*Vasconcella longiflora* (V. Badillo) V. Badillo]
Prim. for., 1900–2050 m; shrub to 4 m with green fls.: 29166, 31239, 31600. [Not in CVPE.]

2. *Carica microcarpa* Jacq. [*Vasconcella microcarpa* (Jacq.) A. DC.]
ssp. *heterophylla* (Poepp. & Endl.) V. Badillo
Prim. for., 1650–2000 m; shrub 1–6 m, fls. green, frs. orange: 30388, 30611, 31132; Q 297; T 155 (last two det. V. Badillo). "Col de monte"

3. *Carica pubescens* Lenné & Koch [*Vasconcella cundinamarcensis* V. Badillo]
Sec. for., 1600–2100 m; shrub or small tree 2–5 m; frs. angled, inflated, orange: 31215, 32441, 32779; C 13074. "Babaco"

CARYOPHYLLACEAE (4)

1. *Arenaria lanuginosa* (Michx.) Rohrb.
Sec. for., banks, 1150–1600 m; weak-stemmed herb, fls. white: 27628, 28824, 31698, 32794.

2. *Cerastium* sp.
Banks of Río Pichán, 1900 m; stems spreading, petals white: 32397.

3. *Drymaria cordata* (L.) Willd. ex Schult.
Sec. for., banks, 1150–2100 m; sprawling herb, fls. white: 27385, 27407, 28825, 32404. "Alverjilla"

4. *Drymaria* cf. *ovata* Willd. ex Schult.
Sec. for., 2050–2100 m; stems flaccid, fls. white: 31534.

5. *Stellaria cuspidata* Willd. ex Schltdl.
Upper mont. for., 2750–2800 m; scrambling herb, fls. white: 31532.

6. *Stellaria ovata* Willd. ex Schltdl.
Sec. for., 1200–1675 m: 29040, 30357; Filskov 37077 (det. J. Madsen).

7. *Stellaria* cf. *prostrata* Baldwin
   Mossy banks in sec. for., 2000 m; scrambling herb, fls. white: 31855. [Not in CVPE, perhaps *S. debilis* d'Urv.]

CECROPIACEAE (2) (* det. C. C. Berg)
   Ref.: Berg, C. C. 1978. Taxon 27: 39–44.

1. *Cecropia gabrielis* Cuatrec.
   Prim. for., 1500–1750 m; tree to 30 m: 32949; G 69901*, 73169*; Q 65*.

2. *Cecropia maxima* Snethl.
   Loma Pahuamba, 1750 m; tree 12 m, barren: F 1564*.

3. *Cecropia monostachya* C. C. Berg
   Sec. for., 1200–1500 m; common tree 5–12 m high; spikes solitary: 27296*, 27635*, 27689*, 32925; Z 119. [Although listed in CVPE as a synonym of *Cecropia gabrielis,* the lower elevational plants appear distinct at Maquipucuna.] "Guarumo"

4. *Cecropia* aff. *radlkoferiana* A. G. Richt.
   Sec. for., 1300–1400 m; tree 10 m, not common: 28167*. [Not in CVPE.]

5. *Pourouma minor* Benoist ?
   Sec. for., 1250 m: 29908 [specimen barren & determination dubious].

CELASTRACEAE (2)

1. *Maytenus macrocarpa* (Ruiz & Pav.) Briq.
   Prim. for., 1900–2250 m; tree 5–6 m, fls. green: 30093, 30232.

2. *Perottetia multiflora* Lundell
   Sec. for., 1550 m; arborescent shrub 8 m: 29254.

CHENOPODIACEAE (1)

1. Chenopodium ambrosioides L.
   Roadside weed, 2000 m: 31231. "Bocino", "Paico"

CHLORANTHACEAE (1) (* det. C. Todzia)
Ref.: Todzia, C. 1990. Fl. Ecuador 40: 1–31.

1. *Hedyosmum cuatrecasanum* Occhioni
Sec. for., 1400–2450 m; tree 12–20 m: C 39724; F 1318; Q 46*. "Tarqui"

2. *Heyosmum goudotianum* Solms var. *goudotianum*
Prim. for., 2200–2250 m; tree to 6 m: 28930*.

3. *Hedyosmum racemosum* (Ruiz & Pav.) Don
Sec. for., 1000–2000 m; tree 4–15 m: 27010*, 28353, 28929, 29311*, 32936;
C 12417*; V 12260*, 12312*.

4. *Hedyosmum sprucei* Solms
Loma Pahuamba, 1750 m; tree 15m, odor of menthol: F 1524.

5. *Hedyosmum uniflorum* Todzia
Sec. for., 1200–1700 m; tree 15 m: 28684*; Q 125*; T 182*, 591*. "Guayusa"

CHRYSOBALANACEAE (1)
Ref.: Prance, G. T. 1972. Fl. Neotrop. 9: 1–410.

1. *Licania* cf. *durifolia* Cuatrec.
Sec. for., 1550–1700 m; tree 5 m: 31701; G 73247.

2. *Licania grandibracteata* Prance
Prim. for., 1500–1700 m; tree 15 m: Q 33 (det. G. Prance).

CLETHRACEAE (1) (* det. G. Gustafson)
Ref.: Gustafson, C. 1992. Fl. Ecuador 45: 1–26.

1. *Clethra obovata* (Ruiz & Pav.) Don
Prim. & upper mont. for., 2000–2300 m; tree 8–20 m: 29449, 30518; F 1441,
1462. "Bermejo"

2. *Clethra revoluta* (Ruiz & Pav.) Spreng.
Upper mont. for., 2450–2750 m; tree 8–15 m: 30592*; F 1363.

CLUSIACEAE (7) (* det. X. Buitrón; ** det. B. Hammel; *** det. W. Palacios).
Ref.: D'Arcy, W. G. 1980. Ann. Missouri Bot. Gard. 67: 969–1043.

[Clusiaceae are abundant in Maquipucuna forests, but probably less adequately understood than any other woody family. Most of the determinations cited here are highly provisional.]

1. *Caraipa* sp. ?
   Sec. for., betw. Cerro Negro & Río Alambi, 1335 m; tree 20 m × 70 cm with plank buttresses: 31312. [Genus not in CVPE.]

2. *Chrysochlamys colombiana* (Cuatrec.) Cuatrec.
   Prim. & rip. for., 1900–2300 m; tree 10–30 m, fls. white: 30262, 32009, 32817.

3. *Chrysochlamys* cf. *dependens* Planch. & Triana
   Prim. & sec. for., 1250–1950 m; arborescent shrub or tree to 20 m, without evident latex, fls. white or pink, frs. reddish: 27169, 27990, 28174, 29099, 30344, 31131, 31553, 32352, 32938, 32952; F 1092, 1095; Z 84, 134. "Cascarillon"

4. *Clusia alata* Triana & Planch.
   Prim. & sec. for., 1250–2250 m; tree 5–10 m, leaves coppery beneath, frs. reddish: 27565*, 28905***, 29371, 29348, 31586, 31702, 31710, 32821; C 39742. "Duco", "Matapalo"

5. *Clusia crenata* Cuatrec.
   Sec. & rip. for., 1300–2200 m; tree 6–25 m, latex yellow; petals white, purple within: 31845, 32760; F 1004, 1015, 1048, 1183; C 5923**; P 3587**; V 12295 (det. J. Pipoly); Zak & Jaramillo 3241**. "Matapalo"

6. *Clusia lineata* (Benth.) Planch. & Triana
   Sec. for., 1200–1650 m; arborescent shrub or tree 5–12 m, fls. white or pink: 27022, 27250*, 27974, 30354, 31735; V 12272**; Z 34, 137.

7. *Clusia multiflora* Kunth
   Loma Pahuamba, 1900–1950 m: F 1051.

8. *Clusia sphaerocarpa* Planch. & Triana
   Sec. for., Loma Cachillacta, 1550–1600 m; arborescent shrub 4–5 m: 30431.

9. *Clusia thurifera* Planch. & Triana
    Sec. for., 1500–1700 m; liana: Q 59 (det. J. Pipoly).

10. *Clusia* sp. 1
    Prim. for., 1750 m; tree 10 m: 27804.

11. *Garcinia madruno* (Kunth) Hammel ?
    Sec. for., 1650 m; tree 7 m: 29358.

12. *Hypericum silenoides* Juss.
    Sec. for., 1200–2650 m; common weed, petals yellow: 27846; K 31964, 32014; Filskov 37086 (det. N. Robson).

13. *Tovomita weddelliana* Planch. & Triana ?
    Prim & sec. for., 1550–1700 m; tree to 15 m, stilt roots: C 39730; G 73182.

14. *Vismia baccifera* (L.) Triana & Planch.
    Sec. & trans. for., 1200–1650 m; shrub or tree to 12 m high, lvs. coppery beneath, fls. greenish: 27700, 31669; Z 169.

15. *Vismia* sp. 1
    Sec. for., 1225 m; shrub 4 m, fls. greenish: 31097. "Mancha roba"

COBAEACEAE (1)

1. *Cobaea trianae* Hemsl.
    Sec. for., 1800–2500 m; prostrate or climbing vine, corolla yellowish-green with purple veins: 30096, 30516; V 12336 (det. A. Prather).

CONVOLVULACEAE (2)
    Ref.: Austin, D. F. 1982. Fl. Ecuador 15: 1–98.

1. *Ipomoea alba* L.
    Sec. for., 1150–1250 m; vine with white fls.: 28157, 28314, 28798.

2. *Ipomoea batatas* (L.) Lam.
    Sec. for. & fields, 1200–1250 m; vine spreading from cult., fls. white to pink: 27538, 28311. "Camote"

3. *Maripa* cf. *nicaraguensis* Hemsl.
    Sec. for., 1250 m; flowers fallen on trail; corolla white: 31010.

CORIARIACEAE (1)
Ref.: Skog, L. 1987. Fl. Ecuador 30: 1–7.

1. *Coriaria ruscifolia* L. ssp. *microphylla* (Poir.) L. E. Skog
Upper mont. for., above Río Pichán, 2500 m; spreading shrub 1–1.5 m, fls. greenish to purplish: 28088; Rubio 2415(?). "Piñán", "Shansi"

CORNACEAE (1)
Ref.: Steyermark, J. A., & R. Liesner. 1987. Ann. Missouri Bot. Gard. 74: 429–430.

1. *Cornus peruviana* J. F. Macbr.
Upper mont. for., 1900–2750 m; tree 10–25 m, fls. whitish: 30593, 31578; F 1003, 1197, 1328; K 31969; P 3610; Zak & Jaramillo 3251 (last two det. Z. Murrell). "Palo de rosa"

CRASSULACEAE (3)

1. Bryophyllum pinnatum (Lam.) Oken
Banks, 1400–2000 m; leaves succulent, fls. reddish: 32937; K 98–122.

2. *Crassula connata* (Ruiz & Pav.) A. Berger
Roadside ditches near Quebrada Santa Rosa, 2000 m; common submerged herb: 31638. [Site found destroyed in 1997.]

3. *Echeveria quitensis* (Kunth) Lindl.
Banks, 2300 m; leaves succulent: K 31959.

CUCURBITACEAE (11) (* det. D. Kearns)
Ref.: Jeffrey, C. 1984. Fl. Suriname 5(1): 457–518.

1. *Apodanthera biflora* Cogn. ?
Sec. for., 1850–2100 m; vine with green frs.: 30226, 30497, 31946. [Elevation anomalously high compared to that listed in CVPE.]

2. *Cayaponia* sp.
Sec. for., Cerro Negro, 1900–2000 m; vine with green frs.: 28023*. "Melón de monte"

3. *Cyclanthera pedata* (L.) Schrad.
   Sec. for., 1200–1300 m; vine with yellow fls., green frs.: 27928, 28294.
   "Cochocho"

4. *Gurania macrophylla* Cogn.
   Sec. for., 1250–1675 m; liana with orange fls.: 27325*, 27632*, 31139, 31546,
   31952; K 31798. "Zapallito"

5. *Gurania pedata* Sprague
   Sec. for., 1250–1400 m; vine with orange fls.: 28784*; Z 114.

6. <u>Lagenaria</u> <u>siceraria</u> (Molina) Standl.
   Disturbed areas, 1200–1350 m; vine with white fls.: 28678, 29383, 31020.
   "Calabaza"

7. *Melothria longituba* C. Jeffrey
   Sec. for., 1900–1950 m; vine with green or yellow fls.: 30105*, 31579.

8. *Melothria pendula* L.
   Sec. for., 1150–1300 m; vine with yellow fls.: 28052, 28144, 28261, 28827,
   31722; Z 39, 200.

9. *Posadaea sphaerocarpa* Cogn.
   Sec. for., 1250 m; vine with green frs.: 28549*, 31724.

10. *Psiguria* cf. *triphylla* (Miq.) C. Jeffrey
    Prim. & sec. for., 1300–1750 m; vine with orange fls.: 27397*, 29039*, 30059.
    [Leaves all simple.]

11. *Rytidostylis carthaginensis* (Jacq.) Kuntze
    Prim. & sec. for., 1200–2000 m; vine with green fls., green explosive frs.:
    27077, 27196*, 27857, 28245*, 29263, 30333, 30390, 31135, 31627; Z 209.

12. *Rytidostylis glabra* (Cogn.) Kuntze
    Sec. for., 1300–1400 m; vine with green frs.: 27970*. [Not in CVPE.]

13. *Rytidostylis trianaei* (Cogn.) Kuntze
    Prim. for.?, 1900–2200 m; fls. yellow-green, frs. green: Zak & Jaramillo 3240
    (det. C. Jeffrey).

14. *Sicydium diffusum* Cogn.
Prim. for.; vine with green fls.: Q 299.

15. *Sicyos* cf. *kunthii* Cogn.
Sec. for., 1150–1300 m; vine with greenish fls.: 28315*, 28799, 31544, 31707; Z 182. "Cohombro"

16. *Sicyos* cf. *warmingii* Cogn.
Sec. for., 1900–2000 m; climbing vine: 30520*. [Not in CVPE.]

CUNONIACEAE (1)
Ref.: Harling, G. 1999. Fl. Ecuador 61: 1–74.

1. *Weinmannia* cf. *latifolia* C. Presl
Prim. & sec. for., 1550–2400 m; shrub or tree to 20 (30) m, lvs. simple; fls. yellowish-white: 27834, 29167, 29369, 29378, 31072, 32801; F 1100; G 69928, 73184; P 3608; Q 21, 83. [Very common in our area, although not listed for Pichincha in CVPE. Our specimens match *W. latifolia* in having strigose pubescence on the midrib on both leaf faces, although petioles are only slightly pubescent. The Gentry specimen 73184 was determined by Harling (1998) as *W. balbisiana* Kunth, and Palacios 3608 as *W. macrophylla* Kunth, and Quelal 21, 83 were determined by David Neill (1999) as *W. macrophylla*. Apparently the distinctions between *W. balbisiana*, *W. latifolia*, and *W. macrophylla* need to be reevaluated on the western slopes of the Ecuadorean Andes.] "Matacha"

2. *Weinmannia lentiscifolia* C. Presl
Prim. & sec. for., 1500–2750 m; shrub 1–15 m; leaves pinnate, rachis winged; fls. whitish: 30314, 30523; K 31971. [Our specimens agree with *W. lentiscifolia* in their larger leaves mostly 10–13 cm long, with strigose rachis, and with fruiting pedicels to 4 mm long. F 1460 was determined as *W. pinnata* L., a species with smaller leaves and shorter pedicels.]

DICHAPETALACEAE (2) (* det. R. Liesner)
Ref.: Prance, G. T. 1972. Fl. Neotrop. 10: 1–84.

1. *Dichapetalum* cf. *nervatum* Cuatrec.
Sec. for., 1400–1500 m; tree 5–10 m, frs. red: 27826*, 31177.

2. *Stephanopodium angulatum* (Little) Prance
Sec. for., 1500–2450 m; tree 50 cm dbh, 30 m, fls. cream: F 1082, 1105, 1327; G 73173; Q 74*, 84*, 96 (det. G. Prance).

EBENACEAE (1)

1. *Diospyros* sp.
Sec. for., trail to Cerro Sosa, 1500 m; tree 15 m: 32964.

[Elaeocarpaceae may be represented by *Sloanea multiflora* Karsten, collected between Nono & Mindo at 1300 m, Dodson et al. 15196.]

ERICACEAE (9) (* det. J. Luteyn)
Refs.: Luteyn, J. L. 1985. Publ. Mus. Ecuat. Ci. Nat., n.s. 4: 5–8; 1991. Nordic J. Bot. 11: 623–627; 1996. Fl. Ecuador 54: 1–404. Smith, A. C. 1932. Contr. U. S. Nat. Herb. 28: 311–511.

1. *Cavendishia bracteata* (Ruiz & Pav. ex St. Hil.) Hoerold
Sec. for. & scrub, 1800–2200 m; shrub 1–2 m, bracts & fls. red, corolla with white tip: 28091, 32753; L 13663; Rubio 2421*.

2. *Cavendishia grandifolia* Hoerold
Prim. & sec. for., 1250–1975 m; ep. shrub with pendulous brs. to 3 m, greenish or pinkish fls.: 27495*, 28703*, 28779*, 31044, 31659.

3. *Cavendishia tarapotana* (Meissner) Benth. & Hook. f. var. *gilgiana* (Hoerold) Luteyn
Sec. for., banks, 1200–1550 m; shrub 1–2.5 m, bracts red, corolla white with pink or red tip: 27315*, 27828*, 30291*, 31118. "Zagalita"

4. *Disterigma acuminatum* (Kunth) Nied.
Prim. & upper mont. for., 1700–2730 m; terr. or ep., bush or scrambling shrub 0.5–1.5 m, fls. white or pink at base: 29134*, 29458*, 30432*; H 29523*, 29547*. "Hualicón de árbol"

5. *Disterigma rimbachii* (A. C. Sm.) Luteyn
Sec. for., Loma Pahuamba, 2100 m; subshrub to 0.5 m, fls. pink, frs. purplish: 30082*; K 98-130.

6. *Gaultheria foliolosa* Benth.
Upper mont. for., Cerro Montecristi, 2730 m; subshrub, fls. white: H 29518*.

7. *Gaultheria insipida* Benth.
   Upper mont. for., Cerro Montecristi, 2750–2800 m; subshrub, fls. magenta: 31504.

8. *Macleania bullata* Yeo
   Prim. & upper mont. for., 1700–2400 m; shrub to 1.5 m with red or orange fls.: 29386*, 29470*, 30163*, 30438*, 31255, 32823; F 1513*.

9. *Macleania coccoloboides* A. C. Sm.
   Rip. for., 2000–2100 m; shrub 1 m, corolla pale lavender with green or purple rim: 31256*, 31543. "Hualicón"

9A. *Macleania coccoloboides* A. C. Sm. × *Macleania ericae* Sleumer
   Cliffsides, 1800–2100 m; shrub to 2.5 m: 30081*; Luteyn & Silva 14790.

10. *Macleania cordifolia* Benth.
   Prim. & sec. for., 1800–1900 m; ep. shrub, corolla pink or red with green tip: 27454*, 27747*. "Hualicón"

11. *Macleania ericae* Sleumer
   Sec. for., 1200–1800 m; ep. or terr., to 2 m, corolla red with green tip, frs. translucent white: 27367*, 27709*, 28963*, 30285*, 30328*, 30418*, 31026, 31893; Boeke 2341; Luteyn 6536; Luteyn & Silva 14789 [neotype].

   [*Macleania loesneriana* Hoerold was described from the "Nanegal Valley", Sodiro 92/2c, [lectotype NY]; this species occurs at higher elevations than most of Maquipucuna, but might be found on Cerro Montecristi.]

12. *Macleania macrantha* Benth.
   Upper mont. for., Cerro Montecristi, 2150–2800 m; shrub to 2 m, sometimes ep., fls. pendent, corolla red with green tip: 29484*, 29491*, 29500*, 29512*, 30526*, 30557*, 31503; C 7188*; F 1293*.

13. *Macleania recumbens* A. C. Sm.
   Prim. for. & trans. to sec., 1300–2250 m; terr. or ep. shrub to 2 m, corolla red with green lobes: 27583*, 29127*, 29344A*, 29420*, 29430*, 29442*, 31048. [25783 & 31048, at lower elevations, 1300–1465 m, have more oblong leaves and resemble *M. ericae* in some respects.]

14. *Pernettya prostrata* (Cav.) DC.
    Upper mont. for., Cerro Montecristi, 2700–2750 m; shrub to 1 m, fls. white: K 32005. "Taglli"

15. *Psammisia aberrans* A. C. Sm.
    Sec. for., 1550–1800 m; shrub 2 m, sometimes climbing, corolla dark red with green lobes, frs. pink: 29344*; C 39740; G 73125*; Luteyn & Silva 14788*; Luteyn & Lebrón-Luteyn 6535.

16. *Psammisia debilis* Sleumer var. *debilis*
    Prim. for., 2000–2700 m; climber or epiphyte, calyx red or orange, corolla white below, red to greenish above: 30151*; K 31982.

16A. *Psammisia debilis* var. *ecuadorensis* Luteyn
    Sec. for., 1900–2400 m: Luteyn & Tirira 13335.

17. *Psammisia ecuadorensis* Hoerold
    Sec. for., 1350–1900 m or more; ep., corolla red, hypanthium greenish in fr.: 29400, 30282*, 30292*; Luteyn & Tirira 13329.

18. *Psammisia sodiroi* Hoerold var. *sodiroi*
    Prim., sec., & upper mont. for., 1600–2800 m; shrub 2 m, corolla red or purplish with green tip: 27127*, 29242*, 29425*, 29475*, 30374*, 30415, 30543*, 31060*, 31099*, 31197*, 31506, 32791, 32829; F 1410; H 29546; Harling & Andersson 11624; Luteyn & Tirira 13331; Sodiro 92/4b [original type coll., cited by Luteyn, 1996]. "Hualicón"

19. *Psammisia ulbrichiana* Hoerold
    Prim., sec., & upper mont. for., 1300–2700 m; common ep. shrub to 3 m, corolla red at base, white distally: 27103*, 27290*, 27489*, 27809*, 28322*, 28708*, 28722*, 29045*, 29143*, 29298*, 29489*, 30283*, 30531*, 31911, 32365; C 39755; G 69923*, 73130*, 73180*; Q 115*; T 168*.

20. *Sphyrospermum buxifolium* Poepp.
    Prim. & sec. for., 1200–2075 m; ep. or terr., often scandent, fls. whitish or pinkish: 27312, 27455, 27678, 27953, 28040*, 28067*, 28231*, 29210, 31561; Boeke 2339*.

21. *Sphyrospermum cordifolium* Benth.
    Prim. & sec. for., 1550–2175 m; terr. shrub 0.5 m, fls. & berries white: 30466* 31337(?); D 6121; T 164*; Croat 72880; Filskov 37112*.

22. *Sphyrospermum grandifolium* (Hoerold) A. C. Sm.
Upper mont. for., 2280–2800 m; ep., fls. red: 29148*, 29492*, 31519.

23. *Sphyrospermum sodiroi* (Hoerold) A. C. Sm.
Prim. for., 1925–2200 m; ep.: 29192*; Harling & Andersson 11612*; Molau & Eriksen 3043*.

24. *Themistoclesia dependens* (Benth.) A. C. Sm.
Upper mont. for., Cerro Montecristi, 2375–2800 m; ep. or vine, fls. red, on long pendulous branches: 29463*, 30559*, 31501, 31505; H 29517*; K 31975; Luteyn & Tirira 13334*.

25. *Thibaudia floribunda* Kunth
Upper mont. for., Cerro Montecristi, 2600–2750 m; shrub or tree to 8 m, fls. pink: 29483*, 30529*, 30551*; H 29522*. "Hualicón de árbol"

26. *Thibaudia martiniana* A. C. Sm.
Sec. for., 1800–2100 m; shrub 1–2.5 m, calyx red, corolla white: 28001*, 30498*.

ESCALLONIACEAE (1)
Ref.: Sleumer, H. 1968. Verh. Konigl. Ned. Akad. Wetensch., Afd. Natuurk., II. 58: 1–146.

1. *Escallonia myrtilloides* L. f. var. *patens* (Ruiz & Pav.) Sleumer
Upper mont. for., Cerro Montecristi, 2725 m; subshrub, fls. white: H 29556.

2. *Escallonia paniculata* (Ruiz & Pav.) Roem. & Schult.
Sec. for., 1675–2200 m; tree 3–11 m, fls. white: 30452, 31953; F 1463.

EUPHORBIACEAE (11) (* det. G. L. Webster; ** det. B. Smith)
Ref.: Webster, G. L., & M. Huft. 1988. Ann. Missouri Bot. Gard. 75: 1087–1144.

1. *Acalypha andina* Müll. Arg.
Rip. & sec. for., 2000–2350 m; shrub 1.5–2 m: 31850; K 32021.

2. *Acalypha dictyoneura* Müll. Arg.
Prim. & sec. for., 1600–2050 m; shrub 1.5–5 m, somewhat scandent: 30475, 31628, 31665, 31939. [A probable synonym is *A. stellipila* Pax & K. Hoffm., described from just outside our area (Gualea), and referred to a different section because of the production of terminal ? spikes.]

3. *Acalypha diversifolia* Jacq.
Sec. for., 1200–1750 m; common weedy shrub 2–4 m: 27186, 27189, 28121; C 6868; F 1252, 1518; V 12326; Z 95, 208. "Pigua", "Moquillo"

4. *Acalypha platyphylla* Müll. Arg.
Prim. & sec. for., 1250–2030 m; common shrub 1.5–3 m: 27265, 27280, 27664, 27741, 27935, 27985, 28339, 28738, 28797, 29087, 30053, 30195, 30235, 30269, 31043, 31944, 32371; C 5919; F 1027*, 1029*, 1088*; V 12330.

5. *Alchornea coelophylla* Pax & K. Hoffm.
Sec. for., 2100 m; tree 25 m: P 3576 (det. R. Liesner).

6. *Alchornea sodiroi* Pax & K. Hoffm.
Rip. woods, 1300 m; bushy shrub or tree 4–5 m: 30445. [The distinctive leaves (tomentose beneath) closely match the description of Pax & K. Hoffmann, based on a Sodiro specimen from the Nanegal Valley.]

7. *Alchornea triplinervia* (Spreng.) Müll. Arg.
Sec. for., 1700–2200 m; shrub or tree 5–15 m: 31207; Q 11; Zak & Jaramillo 3230.

[A remarkable lacuna in our Euphorbiaceae is the complete absence of collections documenting any species of *Chamaesyce*. There is one record of *Chamaesyce jamesonii* (Boiss.) G. L. Webster from the Reserva Geobotanica Pululahua at 3100 m, 78.30 W, Ceron & Montesdeoca 7662. Although this is not close to our eastern boundary, it is possible this species may be found within the limits of the Maquipucuna area.]

8. *Croton coriaceus* Kunth
Rip. for., Quebrada Santa Rosa, 1900–2000 m; tree 5 m, with clear sap: 32407**.

9. *Croton magdalenensis* Müll. Arg.
   Prim. & sec. for., 1250–2200 m; tree 12–20 m, with reddish sap: 32385**, 32816**; Q 135; T 188; B. Smith 306, 307; Zak & Jaramillo 3213. "Sangre de drago"

10. *Croton pavonis* Müll. Arg.
    Sec. for., 1200–1500 m; shrub or tree 3–8 m, with clear sap: 27012, 28143, 29071; V 12259.

    [Euphorbia cotinifolia L., "Barrabas", is commonly cult. along hedgerows.]

11. *Euphorbia laurifolia* Lam.
    Hedgerows, 1400–1700 m; tree or shrub to 4 m, cyathia green: 32773. [Native to Ecuador but plants cultivated where observed in our area.] "Lechero"

12. *Hyeronima fendleri* Briq.
    Sec. for., 1300–1550 m; tree 7–30 m, fls. yellow-white, frs. reddish: 30407, 31036, 31113, 31879. [G 69931 & G 73224 from Cerra de Sosa probably represent the same species.] "Motilón"

13. *Hyeronima macrocarpa* Müll. Arg.
    Sec. for., 1800–2250 m; tree 7 m, frs. red: 31209, 32832; C 7187 (det. R. Liesner); F 1034*, 1055*. "Motilón"

14. *Hyeronima scabrida* (Tul.) Müll. Arg.
    Sec. for., 1750 m; tree 20 m: 30022.

15. Manihot esculenta Crantz
    Roadsides, escaped from cult., 1150–1200 m: 28832. "Yuca"

16. *Margaritaria nobilis* L. f.
    Sec. for., Armenia to Loma San José, 1900–2200 m; tree 4 m: Zak & Jaramillo 3218. "Pepita de puerco"

17. *Phyllanthus anisolobus* Müll. Arg.
    Sec. for., 1250–1325 m; treelet or shrub 2–3.5 m, fls. greenish-white: 31648, 31881, 32904, 32940. "Barbasco"

18. *Phyllanthus niruri* L.
    Sec. for., 1150–1800 m; annual herb: 27011, 28822, 32792; C 12431; Z 52.

19. *Phyllanthus* cf. *symphoricarpoides* Kunth
   Prim. & sec. for., 2200–2450 m; scandent shrub: 32834; F 1365(?).

20. *Phyllanthus sponiifolius* Müll. Arg.
   Sec. for., 1250–1800 m; clambering shrub to 3 m: 27121, 27128, 27738,
   27776, 27812, 27883, 27927, 28743, 29279, 29977, 32375; C 12418, 13008,
   39733; T 191.

21. Ricinus communis L.
   Sec. for. & scrub, disturbed areas, 1200 m: 28676. "Higuerilla"

22. *Sapium glandulosum* (L.) Morong (s. lat.)
   Prim. & sec. for., 1350–1750 m; tree to 10 m, latex white: 29248, 31178;
   C 39739; F 1505; Z 83. [Our specimens agree with *Sapium pavonianum*
   (Müll. Arg.) Huber, as interpreted by Pax & K. Hoffmann; however, that
   species is treated as a synonym in Kruijt's extremely broad circumscription
   (Biblioth. Bot. 146, 1996), which is tentatively adopted here.]

23. *Sapium laurifolium* (A. Rich.) Griseb.
   Loma Pahuamba, 1750 m; tree 7 m: F 1531*.

24. *Sapium stylare* Müll. Arg.
   Prim. & sec. for., 1350–2300 m; tree 12–18 m, latex white: 30234; C 5930
   (det. R. Liesner); F 1419; Z 152. [It is possible that the local tree represents
   S. verum Hemsl., but the distinctions between these two species remain to be
   elucidated]. "Cauchillo"

25. *Tetrorchidium andinum* Müll. Arg.
   Prim. & sec. for., 1250–2000 m; tree or shrub 4–15 m high: 28236, 29101,
   31005, 31206, 32934; C 5921, 12436; F 1537*; G 73230; N 9796; Q 28, 30,
   92, 112; V 12252; Z 98, 221.

FABACEAE (20) (* det. D. Neill; ** det. T. Pennington)
   Ref.: Macbride, J. F. 1943. Publ. Field Mus. Nat. Hist., Bot. 13(3): 3–507.

1. *Calliandra pittieri* Standl. var. *pittieri*
   Sec. for., 1125–1600 m; tree 7–10 m, filaments red at tips: 28356*, 28398,
   31278 (det. R. Barneby), 32796.

2. *Cojoba arborea* (L.) Britton & Rose [*Pithecellobium arboreum* (L.) Urb.]
   Loma Pahuamba, 1750 m; tree 20 m: F 1274*.

3. *Crotalaria nitens* Kunth
   Sec. for. & roadsides, 1200–1250 m; weedy herb to 1 m, fls. yellow: 27305,
   32943.

4. *Crotalaria sagittalis* L.
   Sec. for. & roadsides, 1200–1500 m; weedy herb with yellow fls.: 28399,
   29034.

5. *Dalea coerulea* (L. f.) Schinz & Thell.
   Upper mont. scrub, above Río Pichán, 2500 m; shrub 1 m, fls. purplish: 28089.
   "Izo"

6. *Desmodium adscendens* (Sw.) DC.
   Sec. for., 1200–1400 m; common herb with white fls.: 27589, 27699; Z 21;
   Filskov 37061. "30 reales", "Hierba del dedo"

7. *Desmodium axillare* (Sw.) DC.
   Sec. for., 1200–1400 m; herb with purplish fls.: 27303; Filskov 37085. "Pega-
   pega"

8. *Desmodium campyloclados* Hemsl.
   Sec. for., 1200–2000 m; vine with white fls.: 27365, 28675, 31647, 32932;
   L 14037.

9. *Desmodium incanum* DC.
   Sec. for., 1300–2150 m; vine with pink fls.: 27435, 27566; Z 65.

10. *Desmodium intortum* (Mill.) Urb.
    Clearings, 1100–1500 m; sprawling herb, fls. pinkish-purple turning blue-
    green: 31273, 32966.

11. *Desmodium purpusii* Brandegee
    Sec. for., betw. Marianitas & Nanegal, 1150 m; vine with white fls.: 28817
    (det. N. Zaruma).

12. *Desmodium uncinatum* (Jacq.) DC.
    Sec. for., 1450 m; clambering vine, fls. bluish: 32965.

13. *Dussia lehmannii* Harms
    Prim. for., 1500–2450 m; tree 9–10 m: F 1320; G 73181; Q 100. [Records
    require confirmation, as all specimens are barren.]

14. *Erythrina edulis* Triana ex Micheli
    Prim. for., 1500–1950 m; tree 5–7 m, fls. green, frs. elongated: 32947; F 1110*; Q 294; Rubio 2070. [Appearing very similar to *E. megistophylla*, but leaves smaller & fruits very different; Q 294 was determined as *E. schimpfii* Diels, but that is a lower-elevational species with foliage similar to *E. megistophylla*, and larger flowers.] "Poroton"

15. *Erythrina megistophylla* Diels
    Sec. for., 1200–1650 m; common tree 4–10 m, fls. orange to red or greenish, frs. spherical, indehiscent: 27059, 27261, 27516, 27661, 28306*, 28805*, 30284; G 69937*, 73225*; P 12841; V 12302; Z 149; Espinoza 724. "Porotón"

16. *Erythrina* cf. *smithiana* Krukoff
    Sec. for., betw. Marianitas & Nanegal, 1200–1250 m; tree grown as living fence post, fls. red: 28296, 28686. "Caraca"

17. *Inga densiflora* Benth.
    Sec. for., 1200–1250 m; common tree to 15 m, fls. white: 27372, 28274**, 28352**; Z 41, 197. "Guabo"

18. *Inga nobilis* Willd. ssp. *quaternata* (Poepp.) T. D. Penn.
    Prim. & sec. for., 1250–1550 m; tree, fls. white: 29059; G 73172; Z 126, 218. "Cachi"

19. *Inga oerstediana* Benth. ex Seem.
    Sec. for., 1800–1950 m; tree 10 m, fls. white: 31940; F 1069, 1075**, 1106**.

20. *Inga punctata* Willd.
    Rip. for., Río Umachaca, 1250 m; tree 10: 31747*.

21. *Inga* sp. 1
    Sec. for., 1250 m; tree 12 m, fls. white: 31748.

22. Medicago polymorpha L.
    Roadsides, 2000 m; prostrate weedy herb, fls. yellow: 31813. "Alfafilla"

23. Melilotus indica (L.) All.
    Roadsides, 2000 m; erect weedy herb, fls. yellow: 31862. "Trébol"

24. *Mimosa albida* Humb. & Bonpl. ex Willd. var. *willdenowii* (Poir.) Rudd
    Sec. for., 1250–1300 m; common scandent shrub, fls. pink: 27698; Z 67. "Uña de gato"

25. *Mucuna* cf. *elliptica* (Ruiz & Pav.) DC.
    Sec. for., 1200–1500 m; vine with white fls.: 27095, 27621, 28317; Holm-Nielsen 24506.

26. *Otholobium munyense* (J. F. Macbr.) J. W. Grimes
    Upper. mont. scrub, 2500–2700 m; subshrub with purplish fls.: 28086; K 32003. "Trinitaria", "Trinitaria negra"

27. Pachyrhizus tuberosus (Lam.) Spreng.
    Sec. for., 1250 m; vine with blue fls.: 28737, 31002. [Probably escaped from cultivation.] "Jícama"

28. *Phaseolus coccineus* L.
    Sec. for., 1300–2150 m; vine with red fls.: 30117, 31225; Croat 72900; Hurtado 1442. "Fréjol rojo"

29. *Phaseolus polyanthus* Greenman
    Sec. for., 1300–1920 m; vine with white or pink fls.: 27843, 28329 (det. A. Delgado S.), 30120; N 9793; Hurtado 1435.

30. Pueraria phaseoloides (Roxb.) Benth.
    Sec. for., Cerro Negro, 1700–1800 m; vine with purple fls.: 28045, 30462*, 31954.

31. *Senna bacillaris* (L. f.) H. S. Irwin & Barneby [*Cassia bacillaris* L. f.]
    Sec. for., 1250 m; shrub 3 m, fls. creamy yellow: 31084, 31766; Z 106.

32. *Senna occidentalis* (L.) Link [*Cassia occidentalis* L.]
    Sec. thickets, 1200–1225 m; herb 0.5 m, fls. yellow: 31095. "Aya porotillo"

33. Spartium junceum L.
    Roadsides, c. 2000 m; shrub to 1.5 m, fls. yellow: K 32010. "Retama"

34. Trifolium repens L.
    Roadsides, 2000 m; fls. white: 31224 (det. M. Vincent). "Trébol blanco"

35. Vigna unguiculata (L.) Walp. ?
Sec. for. & scrub, 1200–1250 m; clambering vine with white fls., prob. escaped from cult.: 28284.

36. *Vigna* cf. *vexillata* (L.) A. Rich.
Sec. for., El Carmen to Est. Piscicola, 1150–1200 m; vine with yellow fls.: 28300, 28800, 29074.

FLACOURTIACEAE (3) (* det. R. Liesner)
Ref.: Sleumer, H. O. 1980. Fl. Neotrop. Mongr. 22: 1–499.

1. *Banara guianensis* Aubl.
Sec. for., 1200–1500 m; arborescent shrub or tree 3–8 m, fls. whitish or yellow, fragrant: 27057, 27067, 27251, 28342, 28975; C 12434; V 12255 (det. C. Grandez); Z 99. "Guapilte"

2. *Casearia* cf. *aculeata* Jacq.
Sec. for., 1550 m; tree: F 1000; G 73233.

3. *Casearia cajambrensis* Cuatrec.
Prim. & sec. for., 1200–1900 m; tree 7–20 m: G 73158*; Q 93*, 292*; T 179*, 199*. [Not in CVPE.]

4. *Casearia pitumba* Sleumer
Prim. for., Cerro Sosa & Loma Pahuamba, 1700–1950 m; tree to 7 m: C 39765; F 1112.

5. *Casearia* aff. *prunifolia* Kunth
Upper mont. for., Cerro Montecristi, 2700–2800 m; tree 5–7 m, frs. yellowish & edible: 30596, 31518. [*C. mexiae*, described from just outside our limits to the west, differs in distinctly crenate leaves.]

6. *Casearia quinduensis* Tul.
Prim. for.?, 1900–2200 m; shrub 3 m: Zak & Jaramillo 3235*.

7. *Casearia sylvestris* Sw. ?
Sec. vegetation, 1200–1500 m; tree 5–6 m: V 12248*, 12250*.

8. *Xylosma benthamii* (Tul.) Triana & Planch.
Sec. for., 1300–1500 m; shrub 1–2 m, frs. white or orange: 27988, 29985.

9. *Xylosma* sp. 1
    Sec. for., 1650 m; tree 4 m: 29341.

GENTIANACEAE (3) (* det. J. S. Pringle)
    Ref.: Ewan, J. Contr. 1948. U. S. Nat. Herb. 29: 209–249. Pringle, J. S. 1995.
    Fl. Ecuador 53: 1–131.

1. Centaurium erythraea Raf.
    Clearings in sec. scrub, Quebrada Chiquilpe, 2300 m; herb, fls. pink: 32449.

2. *Irlbachia alata* (Aubl.) Maas
    Sec. for. & banks, 1200–1500 m; herb to 1.5 m, fls. yellow-green: 27073,
    28267. "Lengua de gato"

3. *Macrocarpaea macrophylla* (Kunth) Gilg
    Sec. for., 1200–1700 m: T 154 (det. R. Greissl P.). [Not in CVPE, perhaps *M.
    sodiroana* Gilg.]

4. *Macrocarpaea sodiroana* Gilg
    Prim. & sec. for., 1200–2200 m; to 5 m high, fls. greenish-yellow: 27733*,
    28902, 29065, 31122, 31346, 31550; Harling & Andersson 11575*; G 69921.

GESNERIACEAE (9) (* det. L. E. Skog; Mendoza colls. det. L. P. Kvist & L. E.
    Skog)
    Refs.: Wiehler, H. 1982. Selbyana 6: 1–219. Skog, L. E. 1982. Selbyana 7(1):
    94–99; 1987. Opera Bot. 92: 225–233. Skog, L. E. & L. P. Kvist. 2000. Syst.
    Bot. Monogr. 59: 1–118 [*Gasteranthus*].

1. *Alloplectus bolivianus* (Britton) Wiehler
    Prim., sec., & upper mont. for., 1200–2600 m; clambering shrub, sometimes
    ep., fls. yellow or orange: 27130, 27546*, 27864, 28048, 28326*, 29384*,
    30018, 30381*, 30570*; T 195*; Filskov 37071 (det. L. P. Kvist);
    Mendoza 506.

2. *Alloplectus dodsonii* Wiehler
    Prim. for., 2100–2150 m; shrub to 1 m, bracts red, corolla pale yellow: 29172*.

3. *Alloplectus purpureus* L. P. Kvist & L. E. Skog
    Sec. for., 1200–1700 m; ep. shrub, calyx green, corolla yellow: 27021*, 27137,
    27490, 27503*, 27904*, 27976*, 28685*, 28775*, 29309, 30264*, 30427*;
    V 12261*; Mendoza 507. "Hierba negro"

4. *Alloplectus sprucei* (Kuntze) Wiehler
   Prim. for., Loma Pahuamba, 2460 m; shrub 0.5 m, corolla yellow: F 1349.

5. *Alloplectus tenuis* Benth.
   Prim. & sec. for., 1300–1950 m; terr. or ep., calyx yellow, corolla red, fr. yellow: 28724*, 29168*, 29241*, 31596, 32369; F 1030; T 184*.

6. *Alloplectus tetragonoides* Mansf.
   Prim. & sec. for., 1500–2500 m; ep. or terr. shrub to 1.5 m, calyx green or red to purplish, corolla orange to red: 27324*, 28939*, 29158*, 29189*, 29412*, 30107*, 30250, 30479*, 30577*, 31605, 32408, 32814.

7. *Alloplectus* aff. *tetragonus* (Hanst.) Hanst.
   Rip. for., Reserva Pahuama, 1900 m; ep., calyx maroon, corolla scarlet: 31597*.

8. *Alloplectus teuscheri* (Raymond) Wiehler
   Prim. & sec. for., 1200–2300 m; shrub, usually ep. & scandent, to 2 (4) m, calyx reddish or purplish, corolla yellow or orange: 27680*, 27810*, 27836*, 28024*, 28814*, 28883*, 28936, 29199*, 29215*, 29426*, 30103*, 30145*, 30183, 31241*, 31245*, 31591*, 31822*, 31825*, 31827*, 31833*; C 5951*; P 3581*.

9. *Besleria angustiflora* Fritsch
   Prim. & sec. for., 1500–2750 m; shrub or tree to 10 m, fls. orange: 27172*, 27570*, 28892*, 28910*, 29171*, 29459*, 30500*, 30547*, 32808; H 29543*.

10. *Besleria solanoides* Kunth
    Prim. & sec. for., 1500–2150 m; shrub or tree to 8 m, fls. orange: 28692*, 28202, 30058*, 31656; C 39753; F 1291; N 9804; G 73129*, 73153*; Mendoza 508.

11. *Besleria* cf. *tambensis* C. V. Morton
    Sec. upper mont. for., 2700–2750 m; shrub to 1.7 m, corolla orange: K 31994*. [Not in CVPE.]

12. *Capanea affinis* Fritsch
    Prim. for., 1200–1700 m; fls. violet: T 151*.

13. *Columnea angustata* (Wiehler) L. E. Skog
    Loma Pahuamba, 1950–2000 m; herb to 0.4 m, corolla deep pink: F 1193.

14. *Columnea byrsina* (Wiehler) L. P. Kvist & L. E. Skog
    Sec. for., 1500–2000 m; ep. or terr., corolla orange-pink to purple: 27962*,
    31571*, 32379.

15. *Columnea ciliata* (Wiehler) L. P. Kvist & L. E. Skog
    Prim. & sec. for., 1250–2100 m; ep. shrub, sometimes scandent, fls. red or
    purple: 27140*, 27470, 27478*, 27789, 28178, 28776*, 28895*, 28938*,
    29160*, 30364, 31116, 32928; Q 336*; T 193*; Mendoza 505, 512.

16. *Columnea crassicaulis* (Wiehler) L. P. Kvist & L. E. Skog
    Sec. for., 1300 m; ep., fls. yellow: V 12307*.

17. *Columnea eburnea* (Wiehler) L. P. Kvist & L. E. Skog
    Prim. & sec. for., 1200–1750 m; ep. or terr., lvs. red beneath at tip, bracts
    yellow or greenish, corolla white: 27841*, 28277*, 29126*; G 73207*;
    Mendoza 502, 509.

18. *Columnea mastersonii* (Wiehler) L. E. Skog & L. P. Kvist
    Sec. for., 1250–1650 m; ep., lvs. red at tip, bracts whitish or pale yellow,
    corolla white: 27035*, 31063*, 31123*, 31347*.

19. *Columnea medicinalis* (Wiehler) L. E. Skog & L. P. Kvist
    Prim. & sec. for., 1250–2100 m; ep., bracts reddish at base, corolla yellow with
    red lobes: 29193*, 29244*, 30131*, 31684; F 1246; P 3609*.

20. *Columnea ovatifolia* L. P. Kvist & L. E. Skog
    Prim. for., 2000–2500 m; ep., fls. pink or red: 30162*, 30583*.

21. *Columnea picta* H. Karst.
    Sec. for., 1200–2000 m; ep., leaf tip red beneath, bracts yellow with purple
    spots, corolla yellow: 27070*, 27247, 27255, 27399, 27791, 27808, 27852*,
    28053*, 28225*, 29064*, 29195*, 29234, 29555, 30055*, 30327*, 30402*;
    C 5917*; F 1019; N 8659*; Q 129*; Filskov 37053; Mendoza 503. "Punta de
    lanza"

22. *Columnea strigosa* Benth.
    Prim. for., 1500–2500 m; ep. or clambering shrub, fls. pendent, calyx green,
    corolla orange: 28878*, 28918, 29128, 29183, 29198, 29438, 29451, 30379*,
    30582, 31140*, 31191*, 31198*, 31299*; C 5932*; F 1018; G 73202*;
    P 3607, 3619; T 601*.

23. *Columnea tandapiana* (Wiehler) L. E. Skog & L. P. Kvist
    Sec. for. remnants, Nanegal, 1300 m; ep., calyx green, corolla yellow:
    V 12298*.

24. *Diastema affine* Fritsch
    Banks, sec. for., 1300–1500 m; terr., fls. white: 27379*, 27766, 27871*,
    32974.

25. *Diastema racemiferum* Benth.
    Banks, sec. for., 1300–1475 m; terr., fls. white with purple spots: 27176,
    30386, 32972.

26. *Drymonia brochidodroma* Wiehler
    Prim. & sec. for., 1250–2000 m; terr., subshrub 0.5 m, bracts green, corolla
    white, frs. red: 27194*, 27332, 28201*, 28713*, 28768*, 28863, 31556*,
    31743*. "Labios de puta"

27. *Drymonia macrophylla* (Oerst.) H. E. Moore
    Sec. for., 1300 m; ep.: 31172*.

28. *Drymonia urceolata* Wiehler
    Prim. for., 2100 m; shrub, bracts green or reddish: 29155*.

29. *Gasteranthus lateralis* (C. V. Morton) Wiehler
    Prim & sec. for., 1700–2300 m; herb 0.5–1 m, calyx green, corolla orange:
    27134*, 27274*, 30038*, 30147, 30197, 31573, 31589*, 32847; F 1406.
    [Specimens from higher elevations within the listed range were commonly
    determined as *Gasteranthus* aff. *pansamalanus* (J. D. Sm.) Wiehler.]

30. *Gasteranthus oncogastrus* (Hanst.) Wiehler
    Prim. & sec. for., 1200–1750 m; herb or subshrub to 1 m with orange or brick
    red fls.: 27141, 27794*, 28782*, 30331*.

31. *Gasteranthus quitensis* Benth.
    Prim. & sec. for., 1200–2150 m; herb with red fls.: 27125, 27236, 27792*,
    27896*, 28138*, 29236*, 29237*, 30139*, 30265*, 32973; C 12403*; F 1016;
    G 73195*; N 8656*; V 12285*; Dodson 15748*; Filskov 37090 (det. L. P.
    Kvist). "Zapotito rojo"

32. *Gasteranthus* sp. 1
    Sec. for., Cerro Negro, 2000–2100 m; herb, fls. red: 30480*, 30493*.

33. *Heppiella ulmifolia* (Kunth) Hanst.
Prim. & sec. for., 1250–2725 m; herb or shrub, sometimes scandent, fls. red: 27366*, 28348, 29497, 30421*, 30555*; Mendoza 504. "Oquilla"

34. *Kohleria spicata* (Kunth) Oerst.
Mossy banks, El Carmen to Marianitas, 1200–1250 m; corolla red: 27078, 27627A*; C 6867*. "Oreja de conejo"

35. *Kohleria villosa* (Fritsch) Wiehler var. *anisophylla* (Fristsch) L. P. Kvist & L. E. Skog
Prim. & sec. for., 1250–2250 m; corolla red: 28881*, 29069*, 29142*; Mendoza 510.

35A. *Kohleria villosa* (Fritsch) Wiehler var. *villosa*
Sec. for., 1200–1650 m; herb or vine, corolla pinkish to red: 27193*, 27627*, 27882*, 27893*, 29080*, 30037*; Mendoza 501.

GUNNERACEAE (1) (* det. L. E. Mora-Oseja)
Ref.: Mora-Osejo, L. E. 1984. Flora de Colombia 3: 1–178.

1. *Gunnera atropurpurea* L. E. Mora var. *atropurpurea*
Prim. & sec. for., clearings, 2000–2750 m; herb to 2.5 m, lvs. < 1 m across: 29162*, 30499*, 30538*, 31249; H 29536. [Not in CVPE.]

2. *Gunnera brephogea* Linden & André var. *magna* (L. E. Mora) L. E. Mora
Sec. for., clearings, 1600–2100 m; herb to 2 m, lvs. 1 m across: 27153*.

HIPPOCASTANACEAE (1) (* det. G. T. Prance)
Ref.: Hardin, J. W. 1957. Brittonia 9: 175–178.

1. *Billia columbiana* Planch. & Linden
Sec. for., 1700–2050 m; tree to 15 m, calyx green, corolla creamy-white, anthers orange: 29178, 30233 (det. G. Nesom), 30243*, 30515, 31928; F 1013, 1040; Zak & Jaramillo 3222 (det. M. Merello).

HIPPOCRATEACEAE (1)

1. *Salacia* cf. *cordata* (Miers) Mennega
Rip. for. near waterfall, Sendero Unión de los Ríos, 1300–1350 m; tree 10 m: 32380 (det. with aid of C. Cerón).

HYDRANGEACEAE (1) (* det. E. McClintock; ** det. A. Freire-Fierro)
Ref.: McClintock, E. 1957. Proc. Cal. Acad. Sci. IV. 29: 147–256.

1. *Hydrangea peruviana* Moric.
Prim. & sec. for., 1200–1650 m; common liana to 15 m, leaves remotely denticulate, bracts & fls. red: 27062**, 27608**, 28335, 30003, 30392, 31032, 31106; G 69919, 73174**; K 31803; N 8643**; T 207**, 593**.

2. *Hydrangea tarapotensis* Briq.
Prim. for., 1250–2000 m; climber, leaves entire, fls. yellow-green: 30231*, 32386.

ICACINACEAE (1)

1. *Citronella incarum* (J. F. Macbr.) R. A. Howard
Prim. for., 1750–2300 m; tree 15–28 m: 28488 (det. C. Cerón), 28908; F 1226.

LAMIACEAE (6)
Ref.: Epling, C. 1935–37. Repert. Spec. Nov. Beih. 85(1): 1–96; 85(2): 97–192; 85(4): 289–341.

1. *Hyptis atrorubens* Poit.
Disturbed wet areas, 1125–1300 m; herb with white fls.: 31275; Z 12.

2. *Hyptis obtusiflora* C. Presl ex Benth.
Sec. for., 1300–1500 m; herb to 1.5 m, fls. white: 27374, 27419, 29868, 31742; Z 49. "Hierba de chancho"

3. *Hyptis pectinata* Poit.
Sec. for., 1300–1500 m; herb with white fls.: 28272. "Poleo"

4. *Hyptis sidifolia* (L'Hér.) Briq.
Sec. for., 1200–1500 m; herb 1.5 m, fls. lavender: 27082; Z 121.

5. *Lepechinia betonicifolia* (Lam.) Epling
Banks in upper mont. for., 2750–2800 m; shrub 1 m high, fls. white: C 6872 (cult., 1200 m, det. R. M. Harley); Øllgaard 98988. "Matico"

[*Lepechinia bullata* (Kunth) Epling from 00°01′N, 78°34′W, 2500 m, L 13655, is just east of our boundary.]

6. *Minthostachys mollis* (Kunth) Griseb.
    Sec. & upper mont. scrub, 1550–2500 m; herb 1 m, fls. white: 28096, 31915, 32765. "Tipo"

    [Ocimum campechianum Mill., at 1400 m, Filskov 37014 (det. R. M. Harley, is doubtless cultivated.]

7. *Salvia macrophylla* Benth.
    Sec. for., 1200–1700 m; herb with blue fls.: 27075, 31730, 32758; Z 232.

8. *Salvia quitensis* Benth.
    Prim. & sec. for., 1300–2225 m; shrub or climber to 3.5 m, fls. pinkish: 27364, 27591, 30092, 31848; C 12989.

9. *Salvia scutellarioides* Kunth
    Sec. for., 1200–1400 m: Filskov 37045, 37070 (both det. R. M. Harley).

10. *Scutellaria coccinea* Kunth
    Sec. for., 1100–1300 m; herb with purplish fls.: 27264, 27277, 27665, 28830, 29270, 30325; Filskov 37127 (det. A. Paton).

11. *Stachys lamioides* Benth.
    Sec. for., 1150–2000 m; herb with red fls.: 29073, 30318, 32849.

12. *Stachys micheliana* Briq.
    Sec. for., 1300–1400 m; herb with white fls.: 27428, 31882; Z 166. "Pedorrera"

LAURACEAE (8) (* det. H. van der Werff; ** det. J. Rohwer; *** det. F. G. Lorca-Hernández)
    Ref.: Rohwer, J. G. 1986. Mitt. Inst. Allg. Bot. Hamburg 20: 1–278. 1993. Fl. Neotrop. 60: 1–322.

1. *Beilschmiedia alloiphylla* (Rusby) Kosterm.
    Sec. for., 1200 m; tree 25 m, fls. green: V 12256, 12289.

2. *Beilschmiedia costaricensis* (Mez & Pittier) C. K. Allen
    Prim. & sec. for., 1200–1700 m; tree 20 m: T 141; V 13368 (det. Sa. Nishida). [Not in CVPE.]

3. *Beilschmiedia tovarensis* (Meisn.) Sa. Nishida
   Sec. for., 1300 m; tree 25 m, fls. pale yellow: N 9782 (det. Sa. Nishida). [Not in CVPE.]

4. *Caryodaphnopsis theobromifolia* (A. H. Gentry) van der Werff & H. G. Richt.
   Sec. for., Cerro Negro, 1375–1400 m; tree 10 m, foliage with wintergreen odor: 31315*. "Caoba"

5. *Cinnamomum triplinerve* (Ruiz & Pav.) Kosterm.
   Sec. for., 1300–1600 m; tree 12–15 m, fls. white, with foetid odor, cupules red: 31313***; T 590***; V 12315***.

6. *Licaria applanata* van de Werff
   Loma Pahuamba, 1850–1900 m; tree 10 m: F 1001.

7. *Nectandra acutifolia* (Ruiz & Pav.) Mez
   Sec. for., 1200–1700 m; tree 5–25 m, fls. white: 31037*; C 12440*; N 9779*; Q 133*; T 186*; V 12234, 12246. "Canelo", "Pacche"

8. *Nectandra laurel* Nees
   Prim. & sec. for., 1300–1950 m; tree 12–20 m, fls. white: 28753*, 32757*; F 1063 (det. W. Palacios); P 3618*; Q 80*, 87*; T 160*; V 12297, 13388, 13407.

9. *Nectandra longifolia* (Ruiz & Pav.) Nees
   Sec. for., 1300–1500 m; tree 10 m, fls. white: P 3617*.

10. *Nectandra membranacea* (Sw.) Griseb.
    Sec. for., 1300–2200 m; tree 6–20 m, fls. white: 29357*, 31885; C 39772; N 9775*; P 3620**; Rubio 2420*; Zak & Jaramillo 3225**. "Aguacatillo"

11. *Nectandra obtusata* Rohwer
    Prim. & sec. for., 1300–2200 m; tree 12–25 m, fls. white: 30089; F 1256, 1275; G 69920*; V 12296.

12. *Nectandra purpurea* (Ruiz & Pav.) Mez
    Sec. for., 1400 m; tree 6 m, lvs. slightly aromatic, fls. white, slightly fragrant: 31263*.

13. *Ocotea benthamiana* Mez
    Upper mont. for., 2400–2700 m; tree to 15 m: 29471*, 30597*.

14. *Ocotea* cf. *cernua* (Nees) Mez
Rip. for., 1250–1300 m; tree 4–15 m, fls. green, fr. with red cupule: 29057; K 31806; V 12237, 13398.

15. *Ocotea floribunda* (Sw.) Mez
Prim. & sec. for., 1200–1600 m; tree 10–20 m, fls. yellowish-white: T 603*; V 12277.

16. *Ocotea fuliginosa* Sodiro
Cerro Sosa, 1700 m; tree 20 m: C 39743. [Not in CVPE.]

[*Ocotea insularis* (Meisn.) Mez, between Nanegal & Mindo, V 13384, is probably extralimital.]

17. *Ocotea ira* Mez & Pitt., vel aff.
Sec. for., 1200–1500 m; tree 12 m, fls. green: V 12247.

18. *Ocotea oblonga* (Meissner) Mez
Sec. for., 1200 m; tree 15 m, fls. white: V 12279.

19. *Ocotea rugosa* van der Werff
Sec. for., 1600–2500 m; tree 15 m: Q 52*; V 13366.

20. *Ocotea stenoneura* Mez & Pitt.
Sec. for., 1200 m; tree 10–15 m, fls. green, fragrant: V 12269, 12282, 12284.

21. Persea americana Mill.
Sec. for., near Marianitas, 1200 m; tree 15 m: 28679 [apparently introduced]. "Aguacate"

22. *Persea caerulea* (Ruiz & Pav.) Mez
Prim. for., Loma Pahuamba, 2300 m; tree 20 m: F 1456.

23. *Persea mutisii* Kunth
Prim. for., 1700–2500 m; shrub 6–10 m: C 39759; V 12337.

24. *Persea* aff. *pseudofasciculata* Kopp
Cerro Sosa, 1700 m; tree 25 m: C 39750. [Two additional specimens, F 1157 & V 13390, were collected just south of our boundary at 1600–1950 m.]

25. *Persea* sp. 1 (ined.)
    Sec. for., 1650 m; tree 5 m: 29360*.

26. *Pleurothyrium giganthum* van der Werff
    Sec. for., Cerro Santa Lucia, 1500 m; tree 8 m, bark aromatic: 31762* [barren specimen; identification tentative].

[Several additional collections of Lauraceae remain unidentified.]

LECYTHIDACEAE (3) (* det. S. Mori)
Ref.: Prance, G. T., & S. Mori. 1979. Fl. Neotrop. 21: 141–236.

1. *Eschweilera integrifolia* (Ruiz & Pav. ex Miers) R. Knuth
    Prim. & sec. for., 1550–1700 m; tree 10–20 m, fls. pendent, deep pink: 30355*, 30425*; C 39729; G 73246. [According to Dr. Mori (pers. com.), this is an upper elevational record for this species.]

2. *Eschweilera* cf. *caudiculata* R. Knuth
    Prim. for., 1900 m; flowers (on trail) yellowish: 29044. [Another collection, Z 120, from lower sec. for., may possibly represent the same species.]

3. *Grias* cf. *multinervia* Cuatrec.
    Sec. for., 1350–1450 m; sparsely branching tree 8–10 m; petals creamy-white with reddish tinge: 31755. [32357 may represent a sapling of this species.] "Jagua lechosa"

4. *Gustavia* cf. *dodsonii* S. A. Mori
    Prim. & sec. for., 1325–1700 m; tree 4–12 m; fls. fragrant, whitish tinged with pink: C 39769; G 69930; K 31793. [The Gentry specimen has been identified as *G. pubescens* Ruiz & Pav.] "Membrillo"

5. *Gustavia* cf. *longifolia* Poepp. ex Berg
    Sec. for., 1600 m; sparsely branching shrub 2 m, leaves 1.5–2 m long, frs. green with edible pulp: 31778.

6. *Gustavia* sp. 1
    Sec. for., 1875 m; shrub 2.5 m, frs. green: 30389.

LOASACEAE (2)
Ref.: Poston, M. S., & J. W. Nowicke, Syst. Bot. 15: 671–678. 1990.

1. *Klaprothia mentzelioides* Kunth
Sec. for., 1800–2000 m; flaccid weed with white fls.: 30179, 31609, 32851; C 12996.

2. *Nasa triphylla* (Juss.) Weigend [*Loasa triphylla* Juss.]
Prim. & sec. for., 1300–1900 m; herb to 1.5 m with urticating hairs, fls. white: 27609, 29274, 30111; C 12412; Z 212, 233. "Ortiga borracho", "Pringamoza"

LOGANIACEAE (1) (* det. R. Liesner)

1. *Spigelia multispica* Steud.
Rip. for., 1350–1450 m; herb with whitish fls.: 30268.

2. *Spigelia pendunculata* Kunth
Prim. & sec. for., 2000–2100 m; shrubby herb to 1 m, fls. dull pink: 29085, 30078, 31182; C 5904*, 5918*; P 3588*.

LORANTHACEAE (5) (* det. J. Kuijt)
Ref.: Kuijt, J. 1986. Fl. Ecuador 24: 113–197.

[*Dendrophthora chrysostachya* (C. Presl) Urb., collected between Nono & Nanegalito, Laegaard 51814, is only 2.5' south of our area.]

1. *Dendrophthora lindeniana* Tiegh.
Sec. for., 2000 m; ep.: L 14043.

2. *Gaiadendron punctatum* (Ruiz & Pav.) G. Don
Upper mont. for., 2500–2700 m; tree 6 m, frs. greenish-white: 30603. "Matial"

3. *Oryctanthus alveolatus* (Kunth) Kuijt
Sec. for., 1300 m; ep., parasitic on *Allophylus*, fls. green: 29037.

4. *Phoradendron dipterum* Eichler
Nanegal, 1300 m; ep. shrub with red frs.: V 12321*.

5. *Phoradendron parietarioides* Trel.
Sec. for., 1300–2000 m; ep.: C 7173*; V 12322*.

6. *Phoradendron piperoides* (Kunth) Trel.
   Sec. for., 1300–1400 m; ep.: 27253*.

7. *Phoradendron undulatum* (Pohl ex DC.) Eichler
   Sec. for., 1150–1300 m; ep.: 27653*, 31014*, 31261*, 32902.

8. *Struthanthus aequatoris* Kuijt
   Sec. for., 1250–2200 m; ep., stems vine-like, fls. greenish yellow, berries dull
   red: 30484*, 31937; Zak & Jaramillo 3221*.

9. *Struthanthus leptostachyus* (Kunth) G. Don
   Sec. for., 1200–1350 m; ep. shrub or vine, fls. whitish or yellowish: 27310*,
   28123*, 28682*, 29036*, 29117*, 31175*; V 12303*.

LYTHRACEAE (1) (* det. A. Lourteig)
   Ref.: Lourteig, A. 1989. Fl. Ecuador 37: 1–47.

1. *Cuphea racemosa* (L. f.) Spreng.
   Sec. for. & roadsides, 1250–1500 m; weedy herb, fls. in racemes; petals pink
   or lavender: 27202, 27669, 27911; Z 48; Filskov 37054*. "Hierba de toro"

2. *Cuphea strigulosa* Kunth
   Sec. for. & roadsides, 1200–1400 m; weedy herb, fls. axillary, petals pink:
   27094; N 9778; Z 47; Filskov 37058*. "Hierba de toro"

MAGNOLIACEAE (1)
   Ref.: Lozano-Contreras, G. 1983. Fl. Colombia 1: 1–119.

1. *Talauma* cf. *gilbertoi* Lozano
   Prim. & trans. for., Cerro Sosa, 1600–2000 m; tree to 35 m; fls. yellowish-
   white, fragrant: 27337, 29359, 31672, 32946; C 39747, 39761; G 69912. [Not
   listed in CVPE, and occurring at a higher elevation than any of the four species
   of *Talauma* recorded; closest to *T. gilbertoi* of the Western Cordillera of
   Colombia (Risaralda and Valle) in its leaves pubescent beneath and with scar
   1/2 to 2/3 the length of the petiole, and relatively small gynoecium with c. 10
   carpels.]

MALPIGHIACEAE (1) (* det. C. Anderson)
Ref.: Anderson, C. 1987. Contr. Univ. Michigan Herb. 16: 1–48; Syst. Bot. Monogr. 51: 1–313.

1. *Stigmaphyllon bogotense* Triana & Planch.
Prim. & sec. for., 1300–2175 m; vine with yellow fls.: 29106, 30202*, 31114*, 32390; Croat 72860*.

MALVACEAE (5) (* det. P. A. Fryxell)
Ref.: Fryxell, P. A. 1992. Fl. Ecuador 44: 1–141.

[Abutilon pictum (Gillies ex Hook. & Arn.) Walp. is cult. at Hacienda El Carmen, 1250 m: 31176.]

1. *Bastardiopsis myrianthus* (Planch. & Linden) Fuertes & Fryxell
Sec. for., 1200–1550 m; tree 5–20 m, petals yellow: 27629*, 29261, 31655; N 9790*.

[Malvaviscus penduliflorus DC. is cult. at Hacienda El Carmen, 1250 m; shrub with red fls.: Z 122]

2. *Pavonia castaneifolia* A. St.-Hil. in Naudin
Sec. for., 1400–1800 m; common weed with white fls.: 27175; C 12406*.

3. *Pavonia fruticosa* (Miller) Fawc. & Rendle
Overgrown thickets, 1350 m; weed with pink fls.: 31727; Z 33.

4. *Pavonia schiedeana* Steud.
Sec. for., 1200–1400 m: Filskov 37076*.

5. *Sida acuta* Burm. f.
Sec. for., 1300–1350 m; common weed with yellow fls.: 27421. "Escoba", "Escobilla"

6. *Sida poeppigiana* (K. Schum.) Fryxell
Sec. for., 1200–1400 m: Z 20; Filskov 37055*.

7. *Sida rhombifolia* L.
Sec. for. & roadsides, 1200–1250 m; weed with yellow-orange fls.: 27093. "Escoba"

8. *Urena lobata* L.
   Sec. for., 1300–1350 m; subshrub with pink fls.: 27438; Z 25. "Amonan", "Abrojo", "Cadillo"

9. *Wercklea ferox* (Hook. f.) Fryxell
   Sec. for., 1200–1650 m; spiny shrub 3–4 m, calyx red, corolla yellow: 27544, 29349, 31089; N 8644*, 9791. "Ortiguilla"

MARCGRAVIACEAE (3)
Ref.: de Roon, A. C. 1970. Ann. Missouri Bot. Gard. 57: 29–50.

1. *Marcgravia* cf. *brownei* (Triana & Planch.) Krug
   Prim. for., 1750–2250 m; ep. vine, fls. green: 28942, 30021, 30249, 32836. [G 73120, from 1550–1600 m, perhaps belongs here.]

2. *Marcgravia* cf. *coriacea* Vahl
   Loma Pahuamba, 1900–1950 m: F 1054.

3. *Marcgravia* cf. *nervosa* Triana & Planch.
   Prim. for., 2100–2200 m; ep. vine to 12 m, fls. green: 28901, 29181.

4. *Marcgravia* sp. 1
   Sec. for., 1950 m; ep. vine: 28718.

5. *Marcgravia* sp. 2
   Sec. for., Cerro Campana, 1750 m; climbing vine, fls. greenish: 30033.

6. *Marcgraviastrum sodiroi* (Gilg) Bedell
   Cerro Sosa, 1700 m; vine: C 39746.

7. *Sarcopera anomala* (Kunth) Bedell [*Norantea anomala* Kunth]
   Sec. for., 1200–1750 m; common liana, bracts red, fls. greenish: 27388, 28970, 31764; C 39723; F 1273.

8. *Sarcopera* sp. 1
   Prim. for., hemiephytic climber, 1600–1700 m: G 73163, 73177, 73226.

[Several additional collections of Marcgraviaceae remain unidentified.]

MELASTOMATACEAE (12) (* det. J. J. Wurdack; ** det. F. Almeda; *** det.
E. Cotton)
Ref.: Wurdack, J. J. 1980. Fl. Ecuador 13: 1–405.

1. *Aciotis alata* (Beurl.) Almeda
    Sec. for., 1200–1400 m; herb 1 m, fls. white: 27187, 27877; Filskov 37074*.
    [Wurdack cites Asplund 17278 from Nanegal and Sparre 14892 from
    Nanegalito as somewhat atypical collections of this species.]

2. *Arthrostemma ciliatum* Pav. ex D. Don
    Scrub, El Carmen to Marianitas, 1200–1250 m; scrambling, fls. pale pink:
    27299, 31344.

3. *Blakea eriocalyx* Wurdack
    Prim. & sec. for., 1250–1950 m; shrub or tree to 15 m, corolla white or pink:
    27009, 27248, 27880**, 28199, 28331, 28735*, 29257, 30032, 30406, 31076,
    31956; C 12419*, 39738; G 69941*; Q 293*; V 12267*. [Type from Los
    Puentes near Nanegal, Asplund 17244.]

4. *Blakea quadriflora* Gleason
    Prim., sec., & upper mont. for., 1300–2750 m; shrub or tree to 15 m,
    sometimes vine-like, corolla pink: 28697*, 28709*, 28721*, 28907*, 29108,
    29151, 30031, 31774; F 1285, 1382; T 137*; Z 136.

5. *Blakea rotundifolia* D. Don
    Loma Pahuamba, 1750–2200 m; tree 17–25 m: F 1222, 1473***.

6. *Blakea* sp. 1
    Sec. for., Cerro Negro, 1650 m; tree, bracts green, frs. pink: 31327.

7. *Conostegia superba* D. Don ex Naudin
    Sec. rip. for., 1250–1400 m; common tree 5 m, corolla white: Z 101; Harling &
    Andersson 11588.

8. *Graffenrieda cucullata* (Triana) L. O. Williams.
    Sec. for., 1250–1600 m; arborescent shrub or tree 3–12 m, corolla white:
    27036, 27982, 29050**, 32933; V 12266*; Sparre 14874.

9. *Graffenrieda* sp. 1
    Sec. for., 1450–1500 m; tree 4 m: 31895** [assignment provisional due to lack
    of flowers].

10. *Leandra nervosa* (Naudin) Cogn.
    Sec. for., 1750–1850 m; shrub 1.5–4 m: 31630.

11. *Leandra subseriata* (Naudin) Cogn.
    Sec. for., 1700 m; shrub 2–3 m, fls. white: 32775.

12. *Meriania* cf. *acostae* Wurdack
    Prim. for., 1200–1700 m; tree 12 m, fls. red: T 173*; Nanegal, Sodiro Add. 3.

13. *Meriania* aff. *drakei* (Cogn.) Wurdack
    Loma Pahuamba, 1750–2000 m; tree to 25 m: F 1207, 1225***.

14. *Meriania maxima* Markgr.
    Prim. & sec. for., 1325–1925 m; tree 7–25 m, corolla purple: 31038, 31666; F 1062, 1514; Sparre 14878. "Flor de Mayo"

15. *Meriania peltata* Uribe
    Prim. for., 1900–2000 m; tree 10 m × 10 cm, corolla violet:, 30607* (det. confirmed S. Renner). [Not in CVPE.]

16. *Meriania pichinchensis* Wurdack
    Sec. for., 1900–2200 m; shrub 3 m, corolla red: Zak & Jaramillo 3233*.

17. *Meriania tomentosa* (Cogn.) Wurdack
    Prim. for., 1500–2300 m; shrub or tree to 15 m, corolla red or pink: 28221, 28705*, 31077, 31183, 31324, 31683; C 5906**; F 1049, 1101; G 73191***; N 9798**; Q 23*, 113*, 287**; P 3603**. [A specimen from Cerro Campana, 31077, appears similar but has yellow flowers.] "Hualicón cari"

18. *Miconia aeruginosa* Naudin
    Sec. for., 1200–1800 m; shrub 1–4 m, fls. with foetid odor, corolla white: 27061, 27245*, 27712, 32774; C 12404*; Z 237.

19. *Miconia asclepiadea* Triana
    Prim. & sec. for., 1600–2050 m; shrub or small tree 3–4 m, corolla white: 27139*, 27831**, 29962**, 30478; F 1206; Nanegal, Jameson 363.

20. *Miconia aspergillaris* (Bonpl.) Naudin
    Prim. for., 2500 m; shrub 3 m, fls. white: V 12335*.

21. *Miconia asperrima* Triana
Loma Pahuamba, 2150 m; tree 3 m: F 1288.

22. *Miconia brevitheca* Gleason
Sec. for., 1500–1700 m; tree 15 m: Q 20*, 91*, 108*, 283*.

23. *Miconia crinita* Naudin ssp. *australis* Wurdack
Sec. for., 1800–2000 m; subshrub 1 m: C 7182*.

24. *Miconia dapsiliflora* Wurdack
Prim. for., 1500–1700 m; tree 10 m, calyx yellow, corolla pink: Q 288*.

25. *Miconia denticulata* Naudin
Prim. for., 1900–2000 m; shrub 1–2 m, fls. white: 30132**, 32414.

26. *Miconia hymenanthera* Triana
Prim. & upper mont. for., 1900–2750 m; shrub or tree 2–10 m, calyx red, corolla yellow, orange, or red: 28909*, 29133, 29436, 29482*, 30181, 30534, 31511, 31536, 31610*, 32855.

27. *Miconia lasiocalyx* Cogn.
Prim. for., 1900–2375 m; tree 4–8 m, corolla white: 29456, 29464*, 30178, 30236**.

28. *Miconia rivetii* Dangay & Cherm.
Prim. & sec. for., 1200–2050 m; tree 4–10 m, fls. pale pink, fr. red: 27725*, 29177, 29337, 31716; V 12301*. "Colca"

29. *Miconia sodiroi* Wurdack
Sec. for., Cerro Sosa, 1700 m; tree to 15 m: C 39719; Nanegal, Sodiro Add. 2 [Type collection, cited by Wurdack, 1980].

30. *Miconia theaezans* (Bonpl.) Cogn.
Upper mont. for., 2000–2750 m; shrub or tree 4–12(–30) m, corolla white or pink: 29476*, 30252; F 1494***; H 29525; K 31970; Q 5*, 73* [these two Quelal collections, from 1500–1700 m, are as trees 30 m high, exceeding description of Wurdack]; Zak & Jaramillo 3210*. "Amarillo"

31. *Miconia* sp. 1
Prim. for., 2200–2250 m; shrub 5 m, calyx pink, corolla white: 28941.

32. *Miconia* sp. 2.
    Rip. for., Río Tulambi, 1350–1400 m; shrub 4 m: 30293.

33. *Monochaetum hartwegianum* Naudin
    Prim. for., 2000–2050 m; scarcely woody, fls. pink: 30477. "Amaya"

34. *Monochaetum lineatum* (D. Don) Naudin
    Sec. for., 1800–2000 m; shrub 1 m, fls. pink: Hurtado 1424.

35. *Ossaea micrantha* (Sw.) Macfad. ex Cogn.
    Prim. & sec. for., 1400–2100 m; shrub or tree 1.5–5 m, corolla white: 27184, 27252, 27876**, 28192, 29232, 29376, 30027**, 30200, 31658; F 1007, 1037; G 69926*, 73117***; N 9810**; P 3605*; Q 8*, 90*; Z 108.

36. *Ossaea sparrei* Wurdack
    Sec. for., 1300–1500 m; arborescent shrub 3–4 m, corolla white: 27182*, 27867, 31264.

37. *Tibouchina lepidota* (Bonpl.) Baill.
    Prim. & sec. for., often rip., 1250–2700 m; tree 3–8 m, corolla pink to purple: 27757, 27950, 28388, 30436, 30473, 30600, 31779, 31801, 32761; K 31801. "Flor de Mayo"

38. *Tibouchina longifolia* (Vahl) Baill.
    Sec. for. & scrub, 1200–1350 m; common weedy herb, corolla white: 27081, 27844; D 6115; Z 54; Filskov 37128*.

39. *Tibouchina pendula* Cogn.
    Sec. for., 1500–2250 m; clambering herb or shrub to 1 m, corolla white, frs. reddish: 27954*, 30471, 31957, 32119, 32825.

40. *Topobea pittieri* Cogn.
    Sec. for., 1550–1600 m; shrub or tree or hemiephytic climber 2–5 m, fls. white: 30013, 30429**; G 73126, 73223. [Gentry's specimens recorded as *T. anisophylla* Triana.]

MELIACEAE (3) (* det. W. Palacios)
Ref.: Pennington, T. D., B. T. Styles, & D. A. H. Taylor. 1981. Fl. Neotrop. 28: 1–470.

1. *Carapa* sp.
   Sec. for., northern slopes of Loma Pahuamba, 2025 m: 30091 [barren; identification dubious].

2. *Guarea cartaguenya* Cuatrec.
   Rip. for., 1200–1700 m; tree 30 m: T 181*.

3. *Guarea kunthiana* A. Juss.
   Prim. for., 1500–2450 m; tree to 18 m: F 1345, 1467; G 69927; Q 82; T 175*. "Colorado"

4. *Guarea* sp. 1
   Sec. for., 1650 m; tree 4 m: 29362.

5. *Ruagea glabra* Triana & Planch.
   Prim. for., 1550 m; tree: G 73231*.

6. *Ruagea tomentosa* Cuatrec.
   Prim. for., 1700–2200 m; tree 4–16 m: 30242, 30244; F 1196, 1472; Q 295; P 3590; T 135, 170; Rubio 2419(?).   [Treated as a synonym of *Ruagea pubescens* H. Karst. by Pennington, but appearing distinct in its pointed leaves & pubescent ovary.] "Cedrillo"

MENISPERMACEAE (2)
Ref.: Barneby, R. C. 1970. Mem. N. Y. Bot. Gard. 20(2): 81–158.; Rhodes, D. 1975. Phytologia 30: 415–484 (*Cissampelos*).

1. *Cissampelos grandifolia* Triana & Planch.
   Sec. for., 1200–1300 m; vine with greenish fls. & green frs.: 28281; V 12300 (both det. C. Ott).

2. *Cissampelos tropaeolifolia* DC.
   Prim. & sec. for., 1200–1350 m; common vine with greenish fls.: 27411, 28122, 31179.

3. *Cissampelos* sp. 1

Prim. for., above Río Pichán, 2000–2050 m; vine with green fls.: 30212 [keys to *C. grandifolia* but more robust, with much denser indumentum].

4. *Odontocarya tripetala* Diels

Prim. & sec. for., 1500–2000 m; vine with greenish fls.: 28765, 30347, 32008.

5. *Odontocarya* sp. 1

Sec. for., Inca trail, 1350–1400 m; vine: 32370 [differs from *O. tripetala* in hirsute stems & leaves pubescent beneath].

MONIMIACEAE (1) (* det. S. Renner)

Refs.: Spichiger, R. et al. 1990. Boissiera 44: 1–565. Renner, S. S. & G. Hausner. 1997. Fl. Ecuador 59: 1–125.

1. *Siparuna aspera* (Ruiz & Pav.) A. DC.

Sec. for., 1200–2000 m; common weedy tree 5–13 m, fls. yellow or orange, frs. pinkish with rank odor: 27066, 27304*, 27655, 27731*, 28124*, 28896*, 29093*, 30007, 30362, 31947, 32440; C 5920*; F 1041, 1113, 1248, 1516*, 1557; Q 60*, 64*; T 152*; V 12239*. "Guayusa de monte"

2. *Siparuna echinata* (Kunth) A. DC.

Prim. & upper mont. for., 1900–2750 m; shrub or tree to 5 m: 30255*, 30535; H 29538*; V 12340*, 12342*. "Guayusa", "Limoncillo"

3. *Siparuna lepidota* (Kunth) A. DC.

Sec. & rip. for., 1200–2000 m; shrub or tree 3–6 m, fls. green or yellow: 28806*, 29268*; F 1253; V 12243*. "Guayusa de monte"

4. *Siparuna piloso-lepidota* Heilborn

Upper mont. for., 2375–2500 m; tree to 10 m; fls. pale green: 29468*, 30584; F 1202*, 1431*, 1437; Harling & Andersson 11604*. "Limoncillo", "Palo hediondo"

MORACEAE (6) (* det. C. C. Berg; ** det. T. A. Kvitvik; *** det. W. Palacios)

Refs.: Berg, C. C. 1972. Fl. Neotropica 7: 1–228; 1998. Fl. Ecuador 60: 1–128.

1. *Clarisia biflora* Ruiz & Pav.

Rip. & sec. for., Hacienda El Carmen, 1250–1300 m; tree 15 m: Z 129. "Tillo serrano"

2. *Clarisia* cf. *racemosa* Ruiz & Pav.
   Sec. for., 1250–1300 m; tree 15 m: 31646. "Moral bobo"

3. *Ficus andicola* Standl.
   Sec. for., 1250–1300 m; common tree 5–15 m, figs green with reddish spots:
   27037*; N 9786*.

4. *Ficus brevibracteata* W. C. Burger
   Loma Pahuamba, 1750 m; tree 6 m: F 1585.

   [Ficus carica L., 1400 m; cult.: Filskov 37013]

5. *Ficus cervantesiana* Standl. & L. O. Williams
   Prim. for., 1500–1950 m; tree 12 m, figs pink: 31694; F 1119; Q 79**.

6. *Ficus cuatrecasana* Dugand
   Prim. & sec. for., 1200–2100 m; tree 10–25 m: 30224*, 31018*, 32910;
   F 1501*, 1542*; Q 32**, 57**, 102**; V 12240*.

7. *Ficus dulciaria* Dugand
   Loma Pahuamba, 1550–2470 m; tree 15 m: F 1385*, 1399*, 1488*.

8. *Ficus eximia* Schott ex Spreng.
   Sec. for., 1200–1250 m; tree 20 m, frs. red: 28959.

9. *Ficus* cf. *hartwegii* (Miq.) Miq.
   Sec. for., 1525–1700 m; tree or shrub 4–10 m: 30437*, 31325*.

10. *Ficus macbridei* Standl.
    Sec. for., 1450–1475 m; tree 15 m × 30 cm, leaves with yellowish glands at
    base: 31049*. [Z 105 probably represents the same species.]

11. *Ficus mutisii* Dugand
    Prim & sec. for., 1200–1700 m; tree 7–10 m, frs. yellowish: 27778*, 29379*;
    C 39766.

12. *Ficus rieberiana* C. C. Berg
    Sec. for., 1550 m: G 73216 [paratype].

13. *Ficus* cf. *schippii* Standl.
    Prim. for., 1550 m; tree: G 73128*, 73156*.

14. *Ficus subandina* Dugand
    Prim. & rip. for., 1200–1550 m; hemiepiphytic tree: 29077; G 73244*.

15. *Ficus trigonata* L.
    Prim. for., 1630 m; treelet: G 69932 (det. M. Vázquez Ávila).

16. *Morus insignis* Bureau
    Prim. & sec. for., 1250–2200 m; tree to 15 m: 28337, 29053, 31217; C 12420;
    G 73239; Q 24***, 290***; Z 115; Zak & Jaramillo 3215 (det. R. Liesner).
    "Huasca"

17. *Naucleopsis* sp.
    Rip. for., banks, 2000 m; shrub with milky sap, clambering branches: 31831
    [material barren & verification needed; if correctly placed in *Naucleopsis*, it
    appears distinct from the species treated by Berg (1972) in its long narrow
    leaves with very long petioles].

18. *Pseudolmedia gentryi* C. C. Berg
    Prim. for., 1550 m; tree with white latex: G 73232*.

19. *Pseudolmedia* sp. 1 ?
    Rip. for., 1500 m; tree 3 m: 30001.

20. *Sorocea trophoides* W. C. Burger ssp. *rhodorachis* (Cuatrec.) C. C. Berg
    Prim. for., 1550–1700 m; tree to 6 m: C 39757; G 73121*. "Tillo prieto"

[Myricaceae have not been encountered in Maquipucuna, but *Myrica pubescens*
Humb. & Bonpl. ex Willd. has been collected in Reserva Geobotanica Pululahua,
at 78°30'W, as well as at 00°01'S, 78°41'W, and might possibly occur within our
limits.]

MYRISTICACEAE (1)

1. *Otoba gordoniifolia* (A. DC.) A. H. Gentry
    Prim. & sec. for., 1250–1800 m; common tree to 40 m × 80 cm: 27014, 27681,
    29109, 29119; C 39752; G 69936, 73175, 73255; N 8641, 9807; Q 118; Z 79,
    196. "Caracha coco"

MYRSINACEAE (5) (* det. J. Pipoly)
Refs.: Mez, C. 1902. Pflanzenreich 9 (236): 1–437. Lundell, C. L. 1971. Ann. Missouri Bot. Gard. 58: 285–353. Pipoly, J. J. 1996. Sida 17: 445–458.

1. *Ardisia colombiana* Lundell
Prim. for., 1500–1700 m; tree 12–25 m: Q 13, 69, 72, 76, 86.

2. *Ardisia websteri* Pipoly
Rip. & sec. for., 1250–1675 m; treelet with unbranched trunk 1–3 m, fls. white or reddish, frs. reddish: 27468, 30009*, 30307* [holotype BRIT], 30030, 30361*, 31011, 31323, 32944; T 185*; Z 78, 125. [This recently described species at present is known only from Maquipucuna; according to Pipoly (1996), it is so distinctive that it does not fit easily into any of the subgenera of *Ardisia*.]

3. *Cybianthus goudotiana* (Mez) G. Agostini
Sec. for., 1350–1400 m; shrub or tree: 32366; Z 133.

4. *Cybianthus simplex* (Hook. f.) G. Agostini
Prim. & sec. for., 1250–1950 m; shrub 1.5–2.5 m, fls. pale greenish, frs. reddish to purplish: 27594*, 27795*, 28702*, 28710*, 28769*, 28796*; C 39734.

5. *Geissanthus* cf. *fallenae* Lundell
Prim. for., 1750–2000 m; tree 10 m: 30216; F 1052, 1218, 1233.

6. *Geissanthus* cf. *longistamineus* (A. C. Sm.) Pipoly
Rip. for., 1400 m; tree 15 m, frs. green: 32376.

7. *Geissanthus pichinchae* Mez
Prim. for., 1550 m; treelet: G 73176*. "Casca"

8. *Geissanthus pichinchana* (Lundell) Pipoly
Prim. for., 1550–1900 m; tree to 4 m: F 1008; G 73124*.

9. *Myrsine coriacea* (Sw.) R. Br. ex Roem. & Schult.
Prim. & sec. for., 1200–2200 m; tree to 10 m, fls. green: F 1259; G 73189*; T 176*; Zak & Jaramillo 3244*.

10. *Myrsine* cf. *latifolia* (Ruiz & Pav.) Spreng.
Rip. for., 2000 m; tree 5 m, frs. green: 31844 [differs from description of Mez in more distinctly acuminate leaves not densely puncticulate beneath].

11. *Myrsine pellucida* (Ruiz & Pav.) Spreng.
Sec. for., 175–2100 m; tree 15–20 m, fls. greenish: F 1278; P 3582*.

12. *Stylogyne ambigua* (C. Mart.) Mez
Prim. & upper mont. for., 1700–2720 m; tree to 30 m, fls. white, frs. dark purple or black: 28925, 30248, 32445; C 39754; F 1242, 1310; K 31987; P 3586*.

MYRTACEAE (6) (* det. B. Holst)
Ref.: McVaugh, R. 1958. Publ. Field Mus. Nat. Hist., Bot. 13(4): 567–818.

1. *Calyptranthes* sp. 1 (ined.)
Prim. for., 2050–2150 m; tree to 10 m: 28884*, 29164*.

2. *Calyptranthes* sp. 2
Prim. for., 1900–2000 m; tree 7 m: 30237.

3. *Calyptranthes* sp. 3
Prim. for., 1900–2000 m; tree 12 m: 30238, 31584.

4. *Eugenia* aff. *calva* McVaugh
Sec. for., 1300–1600 m; tree 5 m: 28333*, 29112*; T 592*.

5. *Eugenia* aff. *florida* DC.
Cerro Sosa, 1700 m; tree 8 m: C 39773.

6. *Eugenia* cf. *myrobalana* DC.
Upper mont. for., 2700–2750 m; shrub or tree to 5 m, frs. yellow: 30594; K 31965. [Not in CVPE.]

7. *Eugenia* sp. 1 (ined.)
Prim. for., 2275 m; tree 5 m: 29444*.

8. *Eugenia* sp. 2
Sec. for., 1450 m; tree 30 m: 31112.

9. *Myrcia* cf. *calumbaensis* Kiaersk.
   Prim./sec. for. trans., 1650 m; tree 25 m: 29377. [Not in CVPE.]

10. *Myrcia* cf. *fallax* (Rich.) DC.
    Prim. & sec. for., 1225–1550 m; tree to 5 m: 31040, 31161; G 73170*; Z 132.
    [Leaves are larger than described by McVaugh for Peruvian material.]

11. *Myrcia* sp. 1
    Prim. for., 1850–1925 m; tree 15 m: 28694*.

12. *Myrcia* sp. 2
    Prim. for., 1750–2200 m; tree 17 m: 28754*, 29552*.

13. *Myrcianthes discolor* (Kunth) McVaugh
    Upper mont. for., Cerro Montecristi, 2700 m; tree 15 m: 30595. "Arrayán"
    [Various species of *Myrcianthes* are given this name.]

14. *Myrcianthes orthostemon* (O. Berg) Grifo
    Sec. for., 2075 m; tree 5 m: 31208*.

15. *Myrcianthes rhopaloides* (Kunth) McVaugh
    Loma Pahuamba, 1750–1950 m; tree 30 m: F 1047, 1263.

16. *Myrcianthes* sp.1
    Prim. for., 1550 m; tree: G 73143.

17. Psidium guajava L.
    Sec. for., 1200–1300 m; tree with yellow frs.: 27916; C 6864*; Z 5, 20.
    [Widely planted; although an American species, probably introduced in our
    area.] "Guayaba"

18. Syzygium jambos (L.) Alston
    Sec. for., 1200–1500 m; tree 20 m; fls. pale green: V 12242* [perhaps not
    naturalized].

NYCTAGINACEAE (1)
    Ref.: Bohlin, J.-E. 1988. Nordic J. Bot. 8: 231–252.

1. *Colignonia rufopilosa* Kuntze
    Upper mont. for., 2700–2750 m; scrambling weak shrub to 3 m, bracts white:
    H 29534; K 31974.

OCHNACEAE (1)

1. *Sauvagesia erecta* L.
   Mossy banks in sec. for., 1200–1400 m; herb with white fls.: 27080, 27638; Z 160; Filskov 37066.

OLACACEAE (1)
Ref.: Sleumer, H. 1984. Fl. Neotrop. 38: 1–159.

1. *Heisteria acuminata* (Bonpl.) Engl.
   Sec. for., 1200 m; treelet 6 m, fls. green: V 12275 (det. R. Liesner).

2. *Heisteria* cf. *latifolia* Standl.
   Sec. for., 1200–1250 m; shrub 3 m, fls. yellowish-green: 31163 [perhaps conspecific with the preceding].

ONAGRACEAE (3) (* det. P. Berry)
Ref.: Munz, P. A. 1974. Fl. Ecuador 3: 1–46.

1. *Fuchsia macrostigma* Benth.
   Prim. & sec. for., 1300–2750 m; common shrub to 2 m, fls. pink to red: 27260, 27807, 27961*, 28219, 28220, 28263, 28706, 29169, 30141, 30605*, 31067, 31611, 32377; G 73254*; K 31989; P 3580*. "Zarcillo"

2. *Fuchsia polyantha* Killip ex Munz
   Upper mont. for., 2500–2800 m: L 13653*; Øllgaard 98995 (det. S. León).

3. *Fuchsia scabriuscula* Benth.
   Prim. for., Loma Pahuamba–Tandayapa, 2150–2200 m; shrub 1 m, fls. red & white: 28662*; F 1287.

4. *Fuchsia sessilifolia* Benth.
   Prim. & upper mont. for., 1950–2750 m; shrub 1–3 m, inflors. pendulous, fls. red: 28381, 30087*, 31540, 31581, 32409; L 13646*; Croat 72878*.

5. *Fuchsia sylvatica* Benth.
   Upper mont. for., Cerro Montecristi, 2500–2750 m; shrub c. 1 m, fls. red: 30580*; H 29528*.

6. *Ludwigia affinis* (DC.) H. Hara
   Wet disturbed areas, 1150 m; fls. yellow: 31276.

7. *Ludwigia octovalvis* (Jacq.) P. H. Raven
   Sec. for. & scrub, Marianitas to Nanegal, 1200–1250 m; in wet area: 28308, 31277. "Clavito"

8. Oenothera pubescens Willd. ex Spreng.
   Clearings in upper mont. scrub, 2700 m; herb with reddish-yellow fls.: 31809; K 32002. "Platanillo"

OXALIDACEAE (1)
   Refs.: Lourteig, A. 1980. Ann. Missouri Bot. Gard. 67: 823–850; 1988. Fl. Patagonica 8(5): 1–29.

1. *Oxalis corniculata* L.
   Sec. for., clearings, 2050 m; prostrate herb with yellow fls.: 31203 [possibly not native]. "Chulco"

2. *Oxalis* cf. *frutescens* L.
   Sec. for., 1200–1350 m; herb 0.5 m, fls. yellow: 28937, 31041.

3. *Oxalis jamesonii* Lourteig
   Prim. for.?, 1800 m; fls. yellow: C 12409.

4. *Oxalis* cf. *latifolia* Kunth
   Sec. for., 1250–1450 m; bulbose herb, fls. pink or pale purple: 27402, 29049, 31023; K 31802; Z 176. "Chirisque"

5. *Oxalis lotoides* Kunth
   Sec. for., 1900–2400 m; clambering herbaceous vine, fls. yellow: 31220, 31626, 32805. "Cañitas"

6. *Oxalis psoraleoides* Kunth
   Rip. & sec. for., 1250–1400 m; herb 0.5 m, fls. yellow: 27258, 27598, 27671, 28733, 29089. "Tantanil blanco"

PAPAVERACEAE (1)

1. *Bocconia integrifolia* L.
   Sec. for., 1200–2450 m; shrub to 4 m: 27294, 28851; F 1322; N 9777; Z 180. "Chandor", "Mandor", "Sandalla"

PASSIFLORACEAE (1) (* det. P. M. Jørgensen; ** det. J. MacDougal)
Ref.: Holm-Nielsen, L. B., P. M. Jørgensen, & J. E. Lawesson. 1988. Fl. Ecuador 31: 1–129.

1. *Passiflora adenopoda* DC.
   Sec. for., 2200 m; vine, fls. white with red stamens: MacDougal 4989; Rubio 2418**.

2. *Passiflora alnifolia* Kunth
   Sec. for., 1900–2200 m; vine, fls. purple & cream: Zak & Jaramillo 3232**.

3. *Passiflora chelidonea* Mast.
   Upper mont. for., 2200–2725 m; twining vine, fls. white with purplish-brown markings: H 29521*. "Taxillo"

4. *Passiflora coactilis* (Mast.) Killip
   Prim. for., 21 km west of Calacalí, 2500 m: L 13671*.

5. *Passiflora foetida* L.
   Roadside scrub, 1200–1250 m; vine, leaves foetid, petals white, corona purple: 31721. "Bedoca", "Bombillo"

6. *Passiflora macrophylla* Mast.
   Sec. for., 1250 m; shrub 2.5 m, sparsely branched, fls. white: 31004.

7. *Passiflora manicata* (Juss.) Pers.
   Sec. for., 2050 m; vine, frs. green: 31201*. "Taxo"

8. *Passiflora menispermifolia* Kunth
   Sec. for., banks, 1350 m; vine: 28184.

9. *Passiflora mixta* L. f. var. *eriantha* (Benth.) Killip
   Roadside scrub, 2000 m; vine with pinkish fls.: 31807. "Taxo silvestre"

10. *Passiflora resticulata* Mast. & André
    Sec. for., 1200–2000 m; vine: 27240*, 27547*, 27879*, 28305, 31577; C 5944.

11. *Passiflora riparia* Mart. ex Mast.
    Sec. for., 1400 m; liana, frs. mottled, cut surface with cucurbitacin odor: 31316*.

12. *Passiflora rubra* L.
    Sec. for., 1150–1325 m; vine, frs. red: 31019*, 31271.

13. *Passiflora* sp. nov. ?
    Sec. for., Cerro Negro & Cerro Sosa, 1500–2000 m; nearly leafless vine, calyx green, corolla white: 27958, 32975.

PHYTOLACCACEAE (1)

1. *Phytolacca bogotensis* Kunth
   Clearings in sec. for., 1950 m; herb 0.5 m high, fls. red: 32443.

2. *Phytolacca rivinoides* Kunth & C. D. Bouché
   Sec. for., 1200–2000 m; weed c. 1 (2) m, fls. white or pink, berries bluish: 27471, 27515, 28270, 30121, 31101; Hurtado 1430. "Atusara", "Jaboncillo"

PIPERACEAE (2) (* det. A. J. Bornstein; ** det. R. Callejas)

Refs.: Candolle, C., Bull. Herb. Boiss. 1: 477–495, 505–521. 1898; Sodiro, Piperaceas ecuatorianas. 1900; Tebbs, M. C. Bull. Brit. Mus. Nat. Hist. (Bot.) 19: 117–158. 1989. Trelease, W. & T. G. Yuncker. Piperaceae of Northern South America. 2 vols. 1950.

1. *Peperomia acuminata* Ruiz & Pav.
   Prim. & upper mont. for., 2200–2750 m; robust herb to 1 m, spikes green: 28903*, 29136, 29427, 29466, 29545, 31516.

2. *Peperomia albertsmithii* Trel. & Yunck. var. *villosa* Trel. & Yunck.
   Sec. for., 1650–1675 m; terr., spikes green: 31770**.

3. *Peperomia angularis* C. DC.
   Sec. for., 1250–1950 m; terr. or ep.: 28028**, 28786**, 29221**, 29277**, 31705.

4. *Peperomia bicolor* Sodiro (vel aff.)
   Prim. for., 1900–1925 m: 29194**. [Type collection from Gualea, Sodiro 2/33X.]

5. *Peperomia caespitosa* C. DC.
   Sec. for., 2100 m; ep.: 30083**.

6. *Peperomia calimana* Trel. & Yunck.
Sec. for., 1300–1500 m; ep., spikes green: 27980**.

7. *Peperomia crispa* Sodiro
Sec. for., 1250–1900 m; terr., stems erect or trailing, spikes green: 28265**, 29190**, 31598, 31767, 31892**, 32412, 32971.

[*Peperomia discifolia* Sodiro, from "Mt. Pichincha hasta Gualea", Sodiro s.n., may possibly have been collected within our area.]

8. *Peperomia distachya* (L.) A. Dietr.
Rip. for., 1250–1300 m; ep.: 30512**.

[*Peperomia distichophylla* Sodiro, based on Sodiro 2/5 from Gualea, treated as a synonym of *Peperomia lancifolia* Hook. by Trelease & Yuncker, may occur in our area.]

9. *Peperomia eburnea* Sodiro
Sec. for., 1250–1725 m; very common, terr. or more often ep., mostly creeping or climbing, leaves often red-spotted beneath: 27008**, 27339**, 27704**, 27840**, 27863**, 28072**, 28262**, 28336**, 29252**, 29297**, 29318**, 30008**, 30289, 30326**, 31736, 31926**; Nanegal, 7/1902, Sodiro s.n.

10. *Peperomia ecuadorensis* C. DC.
Sec. for., 1150–1400 m; terr., leaves (at least veins & petioles) reddish beneath: 27179, 27289, 27467, 27686, 27706, 27775**, 28736**, 28818**, 29984; Nanegal–Gualea, Sodiro 4.

11. *Peperomia elongata* Kunth
Nanegal; ep.: 3/1901, Sodiro s.n. [type collection of *Peperomia cuspidigera* Sodiro].

12. *Peperomia emarginella* (Sw. ex Wikstr.) C. DC.
Prim. for., 2050–2100 m; ep.: 29224**.

13. *Peperomia flavescens* C. DC.
Nanegal Valley; ep.: Sodiro. [Not in CVPE.]

14. *Peperomia galioides* Kunth
Sec. for., 1300–2100 m; ep., foliage very aromatic: 29291**, 31533. "Congona silvestre"

15. *Peperomia gaultheriifolia* Sodiro
Nanegal Valley; terr.: 7/1902, Sodiro.

16. *Peperomia haematolepis* Trel.
Nanegal: 10/1901, Sodiro s.n. [type collection of *Peperomia cinerea* Sodiro],
4/1906, Sodiro s.n.

17. *Peperomia hartwegiana* Miq.
Nanegal Valley: 10/1901, Sodiro s.n.

18. *Peperomia helminthostachya* Sodiro
Prim. & sec. for., 1250–1750 m; terr.: 27008a**, 27318*, 27774**. [Cited in
CVPE as known only from the type.]

19. *Peperomia hirtipeduncula* C. DC.
Nanegal Valley: Sodiro 5.

20. *Peperomia longipetiolata* Trel. & Yunck.
Sec. for., 1200–1600 m; terr., creeping: 28750**, 29333**, 31769; Z 97.

21. *Peperomia martiana* Miq.
Sec. for., 1300–1400 m; ep., spikes greenish: 29322**, 31733.

[*Peperomia melanosticta* Sodiro, collected between Nono & Gualea, Sodiro
2/94, may occur in our area.]

22. *Peperomia metapalcoensis* C. DC.
Nanegal Valley: Sodiro s.n. [Not in CVPE.]

23. *Peperomia obtusifolia* (L.) A. Dietr.
Sec. for., 1150–1400 m; terr., spikes greenish: 28292, 28823*, 30334**.

24. *Peperomia pachystachya* C. DC.
Subtrop. for., Nanegal–Gualea: Sodiro 2/15X [type collection of *Peperomia
phyllostachya* Sodiro].

25. *Peperomia peltoidea* Kunth
Rip. for., 1350–1400 m; ep., stems green: 30310**.

26. *Peperomia pilicaulis* C. DC.
Sec. for., 1300–1400 m; ep., climbing: 29281**.

27. *Peperomia pteroneura* C. DC.
    Sec. for., 1300–1400 m; terr., stems prostrate, spikes green: 27760**.

28. *Peperomia* aff. *pubirachis* Yunck.
    Between Nono & Nanegal, 2100 m: Asplund 17269.

    [*Peperomia pseudovariegata* C. DC. var. *sarcophylla* (Sodiro) Trel. & Yunck.,
    based on Sodiro 11/900 from Gualea, may occur in our area.]

29. *Peperomia quadrifolia* (L.) Kunth
    Sec. for., 1200–1600 m; ep.: 27116, 27480, 27701, 27881**, 27906, 29278*,
    29999.

30. *Peperomia rotundata* Kunth var. *rotundata*
    Upper mont. for., 2650 m; ep.: K 31996.

30A. *Peperomia rotundata* Kunth var. *trinervula* (C. DC.) Steyermark
    Prim. for., 2100 m; ep.: P 3596**.

31. *Peperomia rupicola* C. DC.
    Banks in rip. for., 2000 m; terr., stems reddish, spikes green: 31835**.

    [*Peperomia sarmentosa* Sodiro, collected between Mt. Pichincha & Gualea,
    Sodiro 4/900, may occur in our area.]

32. *Peperomia scutellariifolia* Sodiro
    Subtrop. for., Nanegal; prostrate repent herb: Sodiro 2/66X.

33. *Peperomia stelechophila* C. DC.
    Nanegal Valley; repent herb or small scandent shrub: Sodiro s.n. [as
    P. gualeana].

34. *Peperomia subalata* C. DC.
    Nanegal; assurgent herb: 2/1902, Sodiro s.n. [as type collection of *P. camposii*
    Sodiro].

35. *Peperomia subdiscoidea* Sodiro
    "Subtropical forest", Nanegal Valley; repent herb: 1/1900, Sodiro s.n.

36. *Peperomia swartziana* Miq.
   Prim. & sec. for., 1200–2025 m; common ep., sometimes climbing: 27316, 27340, 27694, 27705, 27835, 28013**, 28060, 28208**, 29219, 30267, 30511; V 12328**; Filskov 37103**.

37. *Peperomia ternata* C. DC.
   Prim. & sec. for., 1250–2250 m; terr. or ep., spikes clustered on a peduncle: 29433**, 31557, 32413; Nanegal, Sodiro 7/1900.

38. *Peperomia tetraphylla* (G. Forst.) Hook. & Arn.
   Roadcut in sec. for., 2000 m: K 32012**.

39. *Peperomia tovariana* C. DC.
   Sec. for., 1650 m; ep.: 31866.

40. *Peperomia trinervis* Ruiz & Pav.
   Sec. for., 1400–1800 m; ep.: 32403; 1902, Sodiro s.n.

41. *Peperomia tropeolifolia* Sodiro
   Prim. & sec. for., 2000–2300 m; scandent: 28002, 28247**, 28912**, 29159**, 32830; Nanegal Valley, Sodiro 9/907. "Guiadora"

   [A number of additional collections of *Peperomia* remain unidentified to species.]

42. *Piper aduncum* L.
   Sec. for., 1200–1300 m: N 9776**; T 208**, 209**. "Cordoncillo"

43. *Piper aequale* Vahl
   Sec. for., 1300–2030 m; shrub or tree 2–5 m, foliage very aromatic, spikes erect, whitish or yellowish: 29123**, 31193, 31634, 32415, 32798; K 31909. "Mucuchalla"

44. *Piper appendiculatum* (Benth.) C. DC.
   "Bosques de Nanegal, Hartweg" [Synonymized with *Piper fraseri* C. DC. by Sodiro].

   [*Piper barbatum* Kunth, collected between Calacalí & Yunguillas, Zak & Jaramillo 1809, 1828, probably will be found within our area.]

45. *Piper bullosum* C. DC.
Prim. & sec. for., 1250–2100 m; shrub to 5 m, sometimes clambering or ep.: 27043, 27405, 27978**, 28064**, 28205, 28711*, 29097**, 31237, 31738**, 32372; C 5922**; G 69943**, 73221**; P 3595**; T 161**; Z 102.

46. *Piper carpunya* Ruiz & Pav.
Sec. for., 1500–2200 m; tree or arborescent shrub to 5 m, foliage strongly aromatic, spikes erect, greenish or yellowish: 31909**; Zak & Jaramillo 3209**, 3247** (the last two det. as *Piper lenticellosum* C. DC.).

47. *Piper crassinervium* Kunth
Sec. for., 1300–1500 m; shrub 2–4 m, spikes erect: 27180*, 27991*.

48. *Piper cuspidilimbum* C. DC.
Prim. for., 1630 m: G 69949**.

49. *Piper ecuadorense* Sodiro
Nanegal: 1905, Sodiro s.n.

50. *Piper falanense* Trel. & Yunck.
Sec. for., 1250 m; tree 5 m: 29056**. [Not in CVPE.]

51. *Piper filistilum* C. DC.
Sec. for., 1350–1400 m; shrub 0.5 m: 30315*.

52. *Piper friedrichsthalii* C. DC.
Sec. for., 1350–1550 m; tree 4 m: 29260**; Z 3, 6. [Treated by Bornstein as a synonym of *P. lanceifolium*.]

53. *Piper fuliginosum* Sodiro
Cerro Sosa, 1700 m; tree 4m: C 39743.

54. *Piper grande* Vahl
Sec. for., 1950–2000 m; slender tree 6 m: C 5916**; F 1065.

[*Piper gualeanum* C. DC., based on Sodiro 34 from Gualea, may occur within our area.]

55. *Piper guayasanum* C. DC.
Sec. for., 1300–1680 m; shrub or tree 2.5–5 m, spikes pendent: 27987**, 31338. [Cited in CVPE as known only from the type.]

56. *Piper hispidum* Sw.
    Sec. for., 1200–1700 m; shrub or tree to 8 m, spikes erect: 27041, 29060, 29328**, 32748; T 210**. "Cordoncillo"

57. *Piper hylebates* C. DC.
    Sec. & rip. for., 2000 m; shrub 1.5–2 m, spikes greenish or whitish: 30150*, 31865*.

58. *Piper imperiale* (Miq.) C. DC.
    Nanegal, 1000–1600 m; shrub: Sodiro 1/44.

59. *Piper lanceifolium* Kunth
    Sec. for., 1250–2200 m; arborescent shrub 2–5 m, foliage aromatic, spikes curving at tip: 27038, 30222*, 31232, 31633*; Zak & Jaramillo 3246**. "Cordoncillo"

60. *Piper lineolatifolium* Trel. & Yunck.
    Rip. for., 1350–1400 m; shrub 2 m, spikes erect: 30273*.

61. *Piper manabinum* C. DC.
    Sec. for., 1200–1250 m; shrub: 28288**. "Guabiduca"

62. *Piper marequitense* C. DC.
    Sec. for. & scrub, 1300–1650 m; shrub 3–4 m, spikes erect, yellow: 28283**, 30393*; Z 10. "Luto negro"

63. *Piper nanegalense* Trel. & Yunck.
    Nanegal, 1600 m: Mille s.n., 1908 [holotype ILL]. [Perhaps not separable from *P. obliquum* Ruiz & Pav., fide M. C. Tebb.]

64. *Piper obliquum* Ruiz & Pav. (incl. *P. pseudonobile* C. DC.)
    Sec. for., 1300–2000 m; shrub or tree 2.5–7 m, spikes pendulous, to 1.5 m: 27514, 29366, 30375*, 31890*; F 1024, 1074, 1198; Sodiro 9/899.

65. *Piper ottoniifolium* C. DC.
    Sec. for. along Río Umachaca, 1200 m; ep. shrub, spikes erect, yellowish-green: 27630*.

66. *Piper pallidirameum* C. DC.
    Prim. for., 1800 m; tree 6 m, foliage very aromatic: C 12429. [Not in CVPE.]

67. *Piper peltatum* L. [*Pothomorphe* peltata (L.) Miq.]
Sec. for., 1500 m; herb 1.5 m: 27545. "Corazón", "Santa Maria", "Taco-taco"

68. *Piper phytolaccifolium* Opiz
Sec. for., 1250–1500 m; shrub 1–2 m, spikes erect: 27045, 28116**, 28137**, 28275 (det. N. Greig); V 12238**; Z 8, 19; Filskov 37109**.

69. *Piper piluliferum* Kunth
Sec. for., 1225–1600 m; shrub 1–2 m, spikes capitate: 27396**, 27483**, 27986*, 28687**, 29324**, 30312, 31154; T 211**.

70. *Piper* cf. *scobinifolium* Yunck.
Rip. for., 2000 m; shrub 1 m, spikes erect: 32438. [Not in CVPE.]

71. *Piper sodiroi* C. DC.
Sec. for., 1600–2200 m; shrub or tree 4–5 m; spikes erect, green: 32782; Zak & Jaramillo 3254**.

72. *Piper squamulosum* C. DC.
Sec. for., 1250–1900 m; shrub or tree to 7 m, foliage aromatic, spikes pendulous: 27042*, 29114; Z 29, 192; Nanegal, Sodiro 1/46 [type collection]. "Anis"

73. *Piper subflavispicum* C. DC.
Nanegalito to Armenia & Loma San José, 1900–2200 m; shrub 2 m, inflors. yellow-green: Zak & Jaramillo 3250**.

74. *Piper subglabribracteatum* C. DC.
Nanegal, 1000–1600 m: Sodiro 1/26.

75. *Piper tenuilimbum* C. DC.
Sec. for., 1400–1500 m; shrub: 27983**.

76. *Piper tristemon* C. DC.
Prim. for., 1550 m; treelet, inflors. greenish: G 73211**. [Not in CVPE.]

77. *Piper* aff. *tumidinodum* Yunck.
Sec. for., Cerro Santa Lucia, 1500 m; shrub: 28321**. [Not in CVPE.]

78. *Piper (Pothomorphe) umbellatum* L.
Nanegal: 7/1906, Sodiro s.n. "Maria panga"

79. *Piper* sp. 1 (aff. *bullosum* C DC. ?)
    Sec. for., 1350 m; undershrub 0.5 m: 29991.

[A number of additional collections of *Piper* remain unidentified to species.]

PLANTAGINACEAE (1)

1. Plantago major L.
   Sec. for., weed in clearings, 1400–2700 m: 27429, 28019; K 31980; Filskov 37017. "Llantén"

POLYGALACEAE (3) (* det. B. Eriksen)
Ref.: Ferreyra, R. 1953. Lloydia 16: 193–226.

1. *Monnina cuspidata* Benth.
   Sec. for. & thickets, 1200–1800 m; shrub 2–3 m, fls. purplish: 27246*, 27537*, 27557*, 28056*, 28746*, 28766*, 28767*, 29032; Z 168, 231. "Azulina"

2. *Monnina glaberrima* Chodat
   Prim. for., 1900–2000 m; shrub 4 m, fls. purplish: 30609*. [Not in CVPE.]

3. *Monnina pseudopilosa* Ferreyra
   Prim., sec., & upper mont. for., 1200–2750 m; shrub 1.5–4 m, fls. purplish: 28081*, 28285*, 30599*, 31524, 32809. "Iguilán"

4. *Monnina sodiroana* Chodat
   Prim. & upper mont. for., 1600–2500 m; clambering shrub c. 2 m, fls. purplish: 27275, 27329*, 27679*, 28240*, 28714*, 28760, 28890*, 29188*, 31321, 31664, 31682.

5. *Polygala paniculata* L.
   Sec. for. & clearings, 1250–1600 m; common weed with pinkish fls.: 27644, 32778. "Canchalagua"

6. *Securidaca coriacea* Bonpl. ?
   Sec. for., 1300–1400 m; scandent shrub c. 0.5 m: 32353; "valle Nanegal", Sodiro 59. "Bejuco de jabón"

POLYGONACEAE (2) (* det. J. Brandbyge)
Ref.: Brandbyge, J. 1989. Fl. Ecuador 38: 1–61.

1. *Polygonum hydropiperoides* Michx.
Reported from 10 km north of Nanegalito, 1400–1900 m; weedy herb:
Laegaard 53101*. "Yaco"

2. Rumex obtusifolius L.
Sec. for. & scrub, 1400–1900 m; weedy herb: 31615, 32784; Filskov 37015*.
"Lengua de vaca"

[Anagallis minima (L.) E. H. L. Krause, Primulaceae, from Nono to Nanegalito,
2300 m, Øllgaard 98937, is 2′ south of our boundary.]

PROTEACEAE (2)
Ref.: Sleumer, H. 1954. Bot. Jahrb. Syst. 76: 139–211.

1. *Panopsis antioquensis* L. E. Gut.
Prim. for., 1550 m; tree: G 73140 (det. K. Edwards). [Not in CVPE.]

2. *Panopsis* sp. 1
Prim. & sec. for., 1650–1700 m; tree 10–20 m: 29367; Edwards 565.

3. *Roupala* cf. *obovata* Kunth
Prim. for., 1500–1750 m; tree 5–30 m: 31697; F 1270; Edwards 564, 566, 567.
"Roble andino"

RANUNCULACEAE (3)

1. *Clematis haenkeana* C. Presl
Prim. & sec. for., 1300–2100 m; climbing vine, fls. yellowish-white: 27744,
28185, 30491; Z 17. "Barba de viejo"

2. *Ranunculus limoselloides* Turcz.
Prostrate aquatic herb, sandy banks of stream south of Tandayapa, 1700 m:
32747.

3. *Thalictrum podocarpum* Kunth ex DC.
Rip. for. & banks, 1800–2000 m; fls. greenish: 31837, 32754; P 3622.

ROSACEAE (3) (* det. K. Romoleroux)
Refs.: Romoleroux, Katya. 1994. The family Rosaceae in Ecuador. Dissert., Aarhus. 1996. Fl. Ecuador 56:1–169.

1. *Lachemilla aphanoides* (Mutis ex L. f.) Rothm.
Clearing in upper mont. for., 2725 m: H 29519*; prim. for., 21 km west of Calacalí, 2500 m: L 13664*.

2. *Lachemilla pectinata* (Kunth) Rothm.
Upper mont. for., Cerro Montecristi, 2725–2800 m: 30540, 31521; H 29520*.

3. *Prunus huantensis* Pilg.
Prim. for., 1630 m; tree: G 69940.

4. *Rubus boliviensis* Focke
Sec. for., upper mont. for., 1400–1700, 2700–2750 m; vine with white fls.: 27716*, 29493*, 30536*; C 39731. "Mora" [Name applied to most species of *Rubus*.]

5. *Rubus glabratus* Kunth
Upper mont. for., 2700–2800 m; vine with white fls.: 30537*, 31530.

6. *Rubus glaucus* Benth.
Upper mont. for., 2650–2700 m; vine to 1.2 m, leaves white beneath, fls. white: K 31998.

7. *Rubus nubigenus* Kunth
Upper mont. for., Cerro Montecristi, 2700–2750 m; clambering shrub or vine to 3 m, fls. pink or white: 30548; H 29531*; K 31972. "Huagramora"

8. *Rubus robustus* C. Presl
Loma Pahuamba, 2300 m; spiny shrub, petals pink: F 1428. "Mora silvestre"

9. *Rubus urticifolius* Poir.
Sec. for., 1250–1400 m; vine with white fls.: 27015*, 28319*; Z 167.

10. *Rubus* sp. 1 (ined.)
Sec. for., occ. in prim. for., 1250–1725 m; liana with pink fls.: 27403*, 28351*, 28693*, 29038*, 29284.

RUBIACEAE (22) (revised by P. Delprete; * det. P. Delprete; ** det. C. M. Taylor)
Refs.: Standley, P. C. 1931. Field Mus. Nat. Hist. 7(2): 177–251; Standley, P. C. 1936. Field Mus. Nat. Hist. 13(6): 1–263; Andersson, L. & L. T. Dempster. 1993. Fl. Ecuador 47: 1–38; Andersson, L. & C. M. Taylor. 1994. Fl. Ecuador 50: 1–112; Delprete, P. G., L. Andersson, B. Stahl, & C. M. Taylor. 1999. Fl. Ecuador 62:1–320.

1. *Alibertia hispida* Ducke
Sec. for., 1300–2000 m; tree or shrub 4–6 m: 27993**, 31790; C 5910**. [Not in CVPE.]

2. *Arachnothryx chimboracensis* (Standl.) Steyrm.
Sec. for., 1200–1675 m; shrub to treelet 3–6 m, fls. white to pinkish: 28320*, 28898*, 31015, 31050, 31170, 31772; Q 107*; V 12251 (det. D. Lorence).

3. *Arachnothryx reflexa* (Benth.) Planch.
Sec. for., 1150–1750 m; shrub or tree to 6 m, fls. pink: 27071*, 28841*, 30025*, 31663.

4. *Borojoa stipularis* (Ducke) Cuatrec. ?
Sec. for., 1300–2000 m; tree or shrub 3–12 m: 29987, 30239, 31674, 32956. [It is not certain that the flowers found on the trail, 32954, belong with the other collections.]

5. *Borreria assurgens* (Ruiz & Pav.) Griseb.
Sec. for., 1200–1600 m; weed with white fls.: 27702**, 28149, 32381, 32790; D 6119, 6123; Z 42.

6. *Borreria exilis* L. O. Williams
Sec. for., 1300–1800 m; herb with white fls.: 27368**; C 13021; D 6096; Z 162. "Piojo"

7. *Chiococca alba* (L.) A. Hitchc.
Sec. for., 1900–2000 m; vine: 30614.

8. *Cinchona pubescens* Vahl
Sec. for., 1900–2300 m; tree 15 m: 30613*; F 1454.

[Coffea arabica L., edge of clearings, cultivated, 1200–1250 m, 27539, does not appear to be naturalized.]

9. *Condaminea corymbosa* (Ruiz & Pav.) DC.
    Sec. for., 1150–1550 m; arborescent shrub 3–5 m, fls. yellowish-white to purplish with greenish corolla lobes: 28816, 31563, 31955; D 6104.

10. *Elaeagia utilis* (Goudot) Wedd.
    Prim. for., 1550–1700 m; tree 8 m: C 39721; G 73171 (det. as *E. karstenii* Standl. by C. Taylor).

11. *Faramea calyptrata* C. M. Taylor
    Sec. for., 1900–2200 m; shrub 3 m: C 5914; Zak & Jaramillo 3224**.

12. *Faramea oblongifolia* Standl.
    Prim. & sec. for., 1300–1700 m; shrub or tree 3–5 m, fls. yellowish-white, frs. purplish: 27486**, 28785**, 30019*; D 6085**, G 73116**; N 9805**; Q 1**.

13. *Faramea* cf. *quinqueflora* Poepp.
    Prim. for., 1900–2215 m; tree 7–12 m, fls. blue or blackish: 28949*, 29130*, 29983, 30157, 30261, 30617*.

14. *Faramea* sp. 1 (spp.?)
    Sec. for., 1200–1400 m; shrub 6–7 m: 30416, 31051; D 6070.

15. *Faramea* sp. 2
    Sec. for., 1500–1700 m; shrub 3.5 m, corolla yellow-white: 27136*, 27743*, 27839*, 28077*. "Jazmin"

16. *Galium hypocarpium* (L.) Endl. ex Griseb.
    Prim., sec., & upper mont. for., 1200–2750 m; clambering or creeping herb, corolla greenish, frs. orange: 27092, 27905, 28141, 29537, 30394, 30544; D 6120; K 1963; Z 55; Filskov 37084.   [Sometimes treated as *Relbunium hypocarpium* (L.) Hemsl.] "Coralitos"

17. *Gonzalagunia cornifolia* (Kunth) Standl.
    Sec. for., 1400–1850 m; shrub 1–3 m, fls. & frs. white: 30366*; D 6093.

18. *Gonzalagunia dependens* Ruiz & Pav.
    Sec. for., 1600–2500 m; common shrub 1.5–2 m, fls. white: 27126, 30720, 31094; D 6083; L 13642; Croat 72861; Molau & Eriksen 2189**.

19. *Gonzalagunia discolor* Standl.
    Upper mont. for., banks, 2400 m; undershrub 0.3 m, fls. white: 32804**. [Not in CVPE.]

20. *Hillia parasitica* Jacq.
    Prim. & sec. for., 1200–1400 m; scandent shrub 5–6 m, fls. white: 31712; D 6099.

21. *Hippotis brevipes* Spruce ex K. Schum.
    Sec. for., 1750–1925 m; shrub 2 m, fls. white: F 1043, 1590.

22. *Hoffmannia latifolia* (Bartl. ex DC.) Kuntze
    Sec. for., 1200–2000 m; common cauliflorous shrub or tree 1–2 m, fls. pink or red: 27174*, 27297*, 27973*, 30047, 30135*, 30229, 32378; G 73201**; V 12236**. [Not in CVPE.]

23. *Hoffmannia* cf. *obovata* (Ruiz & Pav.) Standl.
    Prim. for., 1350–2025 m; cauliflorous shrub to 4 m, fls. yellowish to red-orange: 27129*, 28865*, 30270*, 31221, 32364, 32433; D 6094. [Not in CVPE.]

24. *Hoffmannia pauciflora* Standl.
    Prim. & sec. for., 1600 m; shrub 1–2 m, fls. red: D 6081. [Not in CVPE.]

25. *Hoffmannia* sp. 1
    Sec. for., 1200–1400 m; shrub 5–6 m: D 6075.

26. *Hoffmannia* sp. 2
    Rip. for., 1900 m; shrub 2–2.5 m, fls. yellowish-green: 31595.

27. *Ladenbergia pavonii* (Lamb.) Standl.
    Sec. for., 1200–1300 m; tree 5–25 m, fls. white: 27382, 27631 (det. L. Andersson), 28677; N 9774 (det. L. Andersson); Z 147. "Cascarilla"

28. *Manettia acutifolia* Ruiz & Pav.
    Sec. for., 1300–1500 m; vine with white fls.: 27878*. [Not in CVPE.]

29. *Manettia paniculata* Poepp. & Endl.
    Prim. & sec. for., 1450–2000 m; vine with blue or purplish fls.: 27866*, 30251, 31699, 31756.

30. *Manettia* cf. *peruviana* Standl.
    Sec. for., upper mont. for., 1300–2750 m; slender vine with white fls.: 27720;
    K 31968. [Not in CVPE.]

31. *Manettia recurva* Sprague
    Prim. & sec. for., 1500–2750 m; herbaceous vine with white fls.: 27959*,
    28887*, 30563*, 31510, 31938; D 6118; H 29530*; K 31990.

32. *Nertera granadensis* (Mutis ex L. f.) Druce
    Prim. for., 2250–2750 m; mat-forming, berries orange: 29132; K 32001.

33. *Palicourea acetosoides* Wernham
    Prim. for., 1800–2250 m; shrub 3–5 m, inflor. axes yellowish, fls. red:
    27322**, 27651, 28548*, 28876*, 28889*, 28981, 29220, 32828; D 6088.
    "Café de monte" [Application of common name is uncertain.] [Not in CVPE.]

34. *Palicourea amethystina* (Ruiz & Pav.) DC.
    Prim. & sec. for., 1500–2700 m; shrub or tree to 10 m, peduncles lavender, fls.
    blue or purplish: 30517**, 31839; C 5912**; D 6106; F 1092, 1109, 1235;
    K 31966.

35. *Palicourea apicata* Kunth
    Prim. for., 1500 m; shrub 1.5–2m, fls. white: D 6107.

36. *Palicourea calothyrsus* K. Schum. & K. Krause
    Upper mont. for., Cerro Montecristi, 2700–2800 m; shrub 1–1.5 m, fls. white,
    frs. dark blue: 29496**, 30533**, 31522.

37. *Palicourea demissa* Standl.
    Prim. & sec. for., 1450–2250 m; tree or shrub 4–7 m, peduncles orange, fls.
    reddish, frs. bluish-green: 27482**, 27972**, 30095, 31889, 32842; C 39763;
    F 1011, 1563; G 73118**; Q15**, 19**, 38**; Zak & Jaramillo 3219**.

38. *Palicourea guianensis* Aubl.
    Prim. & sec. for., 1300–1750 m; shrub or tree 3–5 m, inflor. axes yellow:
    27131.

39. *Palicourea lineata* Benth.
    Prim. for., 2000–2225 m; tree to 10 m, peduncles pink to lavender, fls. pale
    blue: 28946**, 29174**, 29423**, 30138**.

40. *Palicourea* cf. *padifolia* (Willd. ex Roem. & Schult.) C. M. Taylor & D. Lorence
Prim. for., 1700–1750 m; tree 6 m: 27553. [Not in CVPE.]

41. *Palicourea perquadrangularis* Wernham
Sec. for., Cerro Negro, 1500–2000 m; arborescent shrub, inflor. axes reddish: 28008; G 69911**. [Not in CVPE.]

42. *Palicourea sodiroi* Standl.
Sec. for., 1550–1700 m; tree to 7 m, inf. red: C 39756; G 73241**.

43. *Palicourea stenosepala* Standl.
Sec. for., 17 km NW of Nono: Croat 38865**.

44. *Palicourea thyrsiflora* (Ruiz & Pav.) DC.
Prim. & sec. for., 1500–2150 m; tree 6 m, peduncles orange: D 6068, 6071.

45. *Palicourea* sp. 1
Prim. for., 2300 m; shrub 2.5 m, fls. pink: 29959.

46. *Palicourea* sp. 2
Prim. for., 1900 m; common shrub c. 3 m, inflor. axes orange: 29207.

47. *Pentagonia* cf. *macrophylla* Benth.
Rip. for., 1300–1750 m; tree to 17 m; frs. edible: F 1262; K 31788.

48. *Posoqueria latifolia* (Rudge) Roem. & Schult.
Prim. for., 1900–2000 m; tree 10 m: 30253*; F 1071.

49. *Psychotria allenii* Standl.
Sec. for., 1500–2450 m; tree 7–8 m, fls. yellow-green: 30606**; F 1305; Q 134**; T 187.

50. *Psychotria amplifrons* Standl.
Rip. for., 1250–1300 m: 27659**; Nanegal valley, 1874, Sodiro [type coll.].

51. *Psychotria caerulea* Ruiz & Pav.
Prim. & sec. for., 1300–1915 m; shrub or tree 3–5 m, fls. white: 27133, 28979, 29110, 31604. "Cafecillo" [Name applied to other species of *Psychotria*.]

52. *Psychotria carthagenensis* Jacq.
    Sec. for., 1200–1400 m; shrub 3–4 m, fls. white: D 6077.

53. *Psychotria chimboracensis* Standl.
    Disturbed rip. for., 1300–1440 m; tree 4 m, frs. purplish: 27288**.

54. *Psychotria* aff. *ernestii* Krause
    Cerro Sosa, 1700 m; tree or shrub 3 m: C 39741.

55. *Psychotria gentryi* (Dwyer) C. M. Taylor
    Sec. for., 1250–2000 m; tree or shrub 4–18 m, fls. white: 27596*, 27992*, 29082*, 30023*, 30254, 30256, 31157, 31341, 31653; D 6069; Z 116.

56. *Psychotria hazenii* Standl.
    Prim. & sec. for., 1350–2000 m; arborescent shrub 2–4 m, fls. white: 28216**, 30129**, 30377**; F 1002, 1072; G 73214**; Z 80**; Hurtado 1433**. "Pigüe"

57. *Psychotria lateriflora* Standl.
    Prim. for., 1800 m; shrub, fls. white: 27682**. [Not accepted in CVPE.]

58. *Psychotria longipedunculoides* C. M Taylor
    Prim. for., 1800–2000 m; shrub 2.5 m, fls. white: 27328**. [Not in CVPE.]

59. *Psychotria longissima* Standl.
    Sec. for., 1300 m; shrub 1 m, fls. & frs. white: V 12318**.

60. *Psychotria macrophylla* Ruiz & Pav. (s. lat.)
    Prim. & sec. for., 1250–2000 m; common herb 1–2 m, fls. & frs. white: 27469, 27862, 27901, 28022, 28070, 28209, 28791, 29116, 29226, 30182, 30297; C 7178; D 6072, 6109. [In the narrow circumscription, these collections may represent *Psychotria longissima* Standl.]

61. *Psychotria siggersiana* Standl.
    Sec. for., 1500 m; subshrub 1–1.5 m, fls. white: 27201 (det. J. Dwyer).

62. *Psychotria steyermarkii* Standl.
    Prim. for., 1550–2100 m; shrub 1.5–2 m, fls. white: 28950*; D 6087. [Not in CVPE.]

63. *Psychotria* sp. 1
   Sec. for., 1200–1400 m; shrub 4–5 m, fls. white: D 6067, 6084.

64. *Psychotria* sp. 2
   Sec. for., 1650 m; herb 1 m: 30040.

65. *Psychotria* sp. 3
   Sec. for., 1200–1400 m; shrub 5–6 m: D 6074.

66. *Sabicea villosa* Willd. ex Roem. & Schult.
   Sec. for., 1200–1500 m; vine with white fls.: 27376 (det. L. Andersson), 28120; Z 38; Filskov 37068 (det. L. Andersson).

RUTACEAE (2)
   Ref.: Kaastra, R. C. 1982. Flora Neotropica 33: 1–198.

1. *Esenbeckia warscewiczii* Engl. (vel aff.)
   Sec. for., 1275 m; tree 8 m: 31744 [capsules not tuberculate, nor with apophyses; not conspecific with *E. grandiflora* var. *peruviana* Macbr.; possibly representing *E. dielsiana* Shulze, which was referred to the synonymy of *E. warscewiczii* by Kaastra (1982).]

2. *Zanthoxylum quinduense* Tul.
   Loma Pahuamba, 2450–2500 m; treelet 3 m: F 1374.

SABIACEAE (1)
   Refs.: Cuatrecasas, J., & J. M. Idrobo. 1955. Caldasia 7: 187–211. Gentry, A. H. 1986. Ann. Missouri Bot. Gard. 73: 820–824.

1. *Meliosma arenosa* Idrobo & Cuatrec.
   Loma Pahuamba, 1750 m; tree 18 m: F 1255, 1568.

2. *Meliosma* aff. *occidentalis* Cuatrec.
   Loma Pahuamba, 1900–1950 m; F 1039.

3. *Meliosma* aff. *violacea* Cuatrec. & Idrobo
   Prim. for., 1900–2000 m; tree 5 m: 30240.

4. *Meliosma* sp. 1
   Prim. for., 1550 m; tree: G 73152, 73178.

SAPINDACEAE (6)
Ref.: Croat, T. B. 1976. Ann. Missouri Bot. Gard. 63: 419–540.

1. *Allophylus floribundus* (Poepp.) Radlk.
Sec. for., 1250–1950 m; shrub or tree to 12 m, fls. white, frs. orange: 27064, 28106, 29055, 29113, 29334, 29982, 31348, 31704; F 1090. [27722, recorded as a liana with yellow frs., may represent a different species.]

2. *Cupania cinerea* Poepp.
Prim. for., 1700–1800 m; tree 6–17 m, fls. cream: C 12424, 39726; F 1566.

3. *Matayba* sp.
Prim. for., 1550 m; tree 0.5 m dbh: G 73164.

4. *Paullinia* cf. *fraxinifolia* Triana & Planch.
Sec. for., 1200–1600 m; liana, fls. white: 31267. [Not in CVPE.]

5. *Paullinia gigantea* Poepp.
Prim. & sec. for., 1200–1600 m; liana, frs. reddish: 28392, 28960; Nanegal, Sodiro 279 [syntype of *Paullinia quitensis* Radlk., fide Radlkofer, Bot. Jahrb. 36: 382. 1905].

6. *Serjania mollis* Kunth
Sec. for. & scrub, 2000–2300 m; vine: 31818, K 32019. "Bejuco negro"

7. *Talisia cerasina* (Benth.) Radlk.
Prim. & sec. for., 1225–1750 m; tree 2.5–6 m, leaves to 1 m, petals white or pale pink: 27287, 27556, 30441, 31107, 31888; F 1269. "Paragüita" [Not in CVPE.]

SAPOTACEAE (1)
Ref.: Pennington, T. D. 1990. Fl. Neotropica 52: 1–771.

1. *Pouteria* cf. *collina* (Little) T. D. Penn.
Sec. for., 1300–1700 m; tree 5–15 m, stems with white latex, fls. yellow-green: 30014, 30363, 31749. [31047, with more pointed leaves & longer petioles, may represent another species.]

SCROPHULARIACEAE (9)
Ref.: Holmgren, N. N., & U. Molau. 1984. Fl. Ecuador 21: 1–188.

1. *Alonsoa meridionalis* (L. f.) Kuntze
   Loma Pahuamba, 2300 m; herb, corolla white: F 1249.

2. *Bartsia mutica* (Kunth) Benth.
   Upper mont. scrub, above Río Pichán, 2500 m; herb with pink fls.: 28094.

3. *Calceolaria chelidonioides* Kunth
   Clearings in sec. for., 1950 m; herb with yellow fls.: 32442.

4. *Calceolaria crenata* Lam.
   Sec. for. & upper mont. scrub, above Río Pichán, 1900–2500 m; herb with yellow fls.: 28087, 30123. "Zapatito"

   [*Calceolaria tenuis* Benth., collected along Calacalí–Nanegalito road, 2175 m, Croat 72874 (det. U. Molau), is only 1' east of our boundary.]

5. *Calceolaria perfoliata* L. f.
   Prim. for., 21 km west of Calacalí, 2500 m: L 13643.

6. *Castilleja arvensis* Schltdl. & Cham.
   Prim. & sec. for., 1200–2000 m; herb with pink or red bracts: 27144, 28268, 30124, 30465; Z 28; Filskov 37056 (det. N. H. Holmgren). "Candelilla"

7. *Escobedia grandiflora* (L. f.) Kuntze
   Scrub, El Carmen to Marianitas, 1200–1300 m; herb to 1 m, fls. white: 27090, 27167. "Malva blanca"

8. *Lamourouxia virgata* Kunth
   Upper mont. scrub, 2300–2500 m; herb with pink fls.: 28095; K 31960; Rubio 2416. "Falsa dedalera"

9. *Leucocarpus perfoliatus* (Kunth) Benth.
   Sec. for., 1950–2175 m; herb with yellow fls.: 28080, 30086, 31613, 32405; C 5945; J. F. Smith 1981.

10. *Scoparia dulcis* L.
    Sec. for., 1200–1500 m; weedy herb, fls. white: 27192, 27408, 28154. "Tía tina", "Pedepe"

11. *Stemodia suffruticosa* Kunth
Sec. for., 1875–2100 m; subshrub to 1.5 m, fls. bluish: 30079, 30476, 31632; C 5943 (det. K. Barringer); Hurtado 1429.

SIMAROUBACEAE (1)

1. *Picramnia* sp.
Rip. for., Río Umachaca, 1250 m; tree 8 m: 31649. [Verification of Simaroubaceae in our area is needed, as the single collection is vegetative.]

SOLANACEAE (18) (* det. S. Knapp; ** det. E. Dean; *** det. M. Nee)
Refs.: D'Arcy, W. G. 1973. Ann. Missouri Bot. Gard. 60: 573–780. Knapp, S., et al. 1997. Ann. Missouri Bot. Gard. 84: 67–89; 1998. Novon 8: 152–161.

1. *Acnistus arborescens* (L.) Schltdl.
Sec. for., 1200–1250 m; shrub or tree 5–8 m, fls. white or yellowish, fragrant: 28155, 28282, 31715.

2. *Browallia americana* L.
Sec. for., 1200–1675 m; herb with blue fls.: 27197, 27423, 29048, 30360.

3. *Browallia speciosa* Hook.
Sec. for., 1300–1800 m; herb with blue fls.: 28271, 30743, 31334.

4. *Brugmansia candida* Pers.
Prim. & sec. for., 1200–1800 m; shrub or tree to 6 m, fls. pendent, white: 28222, 28720, 31335**, 31717, 31761. [This plant often goes under the name *Brugmansia suaveolens* (Humb. & Bonpl. ex Willd.) Bercht. & C. Presl, but A. Holguín has determined 28222 as *B. candida*.] "Floripondio blanco"

5. *Brugmansia versicolor* Lagerh.
Prim. for., 1200–1700 m; tree 6 m, fls. white: 28897 (det. A. Holguín); Q 58*; T 144 (det. A. Holguín). [It remains to be demonstrated whether we really have two white-flowered species of *Brugmansia* at Maquipucuna.]

6. *Capsicum geminifolium* (Dammer) Hunz.
Sec. for., 2075–2100 m; shrub 1 m, corolla yellow: 30501 (det. N. W. Sawyer).

7. *Capsicum lycianthoides* Bitter

Prim. & sec. for., 1300–2100 m; shrub 1–2.5 m, fls. yellow, frs. red or orange: 27143***, 27493***, 27602***, 27714**, 27888**, 27984***, 28749***, 28856***, 29094**, 29217**, 29283**, 29345**, 30044***, 32907; P 3591***; Z 82.

8. *Cestrum humboldtii* Francey

Sec. for., 1900–2200 m; shrub 5 m, corolla cream: Zak & Jaramillo 3214***.

9. *Cestrum megalophyllum* Dunal

Sec. for., 1250–2150 m; shrub or tree to 10 m, fls. white or pale yellow, frs. purplish: 27198, 27256, 27633, 27707, 29051, 29096(?), 29098, 31006, 32362; F 1006, 1297, 1528; G 73228; Z 45, 201. "Sauco negro"

10. *Cestrum peruvianum* Hort. Roth. ex Dunal

Upper mont. & rip. for., 2000–2400 m; shrub 2–3 m, frs. purplish-black: 31204***, 31248***, 32812, 32833. "Sauco"

11. *Cuatresia colombiana* Hunz.

Rip. for., Quebrada Santa Rosa, 2000–2050 m; shrub 1.5 m, corolla yellow: 31243**, 31250**, 31832. [Not in CVPE.]

12. *Cuatresia harlingiana* Hunz.

Rip. & sec. for., 1250–1900 m; shrub or subshrub 0.5–2.5 m, corolla white with purple center: 31151**, 31168***, 31688***, 32927; F 1026***.

13. *Cuatresia riparia* (Kunth) Hunz.

Sec. for., 1600 m; shrub 1.5 m, fls. yellow: 30346**, 32957**.

14. *Cuatresia* sp. 1

Rip. for., Quebrada Santa Rosa, 2000 m; shrub 1 m, corolla purple with white border: 31240**.

15. *Cuatresia* sp. 2

Sec. for., 1650 m; shrub 1.5 m, fls. & fr. white: 31133**.

16 *Cyphomandra betacea* (Cav.) Sendtn.

Sec. for., 1400 m: Filskov 37010 (det. L. Bohs). [*Solanum betaceum* Cav. in CVPE.]

17. *Cyphomandra hartwegii* (Miers) Walp.
    Prim. for., 2075–2100 m; arborescent shrub 3 m, petals white with purple midstrip: 30492**. [= *Solanum circinatum* Bohs.] "Tomate de monte"

18. *Cyphomandra hypomalaca* Bitter
    Prim. & sec. for., 1175–2100 m; arborescent shrub 2–3 m, fls. white or greenish tinged with purple: 27585, 27710 (det. L. Bohs), 28810, 31103**; Z 227. [*Solanum fallax* Bohs in CVPE.]

19. *Deprea glabra* (Standl.) Hunziker
    Prim. for., 1800–2000 m; shrub 4 m, fls. blue-green: C 7183***.

20. *Deprea granulosa* (Miers) Hunz. & Barbosa
    Sec. for., Cerro Santa Lucia, 1700 m; herb 1 m, fls. yellow: 28345.

    [*Dunalia solanacea* Kunth, collected at 2700 m near Yunguillas, 31529, and at 2000 m, Dodson & Thien 1081, may occur within our area.]

21. *Iochroma calycinum* Benth.
    Sec. for., 1900–2200 m; shrub 2 m, fls. & frs. violet: Zak & Jaramillo 3248***.

22. *Jaltomata glandulosa* (Miers) R. Castillo & R. E. Schult.
    Roadside, 2000 m; viscid herb 0.5 m, corolla lavender: 32401.

23. *Jaltomata procumbens* (Cav.) J. L. Gentry
    Sec. for., 1150–1750 m; herb c. 1 m, corolla yellow or white with yellow-green spots: 27185, 27387, 27422, 28838, 30062, 31884. "Yana cucuna"

24. *Jaltomata sinuosa* (Miers) Mione
    Rip. for., 2000 m; herb with clammy foliage, corolla whitish with purple center: 31234 (det. T. Mione).

25. *Juanulloa pavonii* (Miers) Benth. & Hook. f. [*Markea pavonii* (Miers) D'Arcy]
    Sec. for., 1200–1750 m; ep. (?) liana or tree with vine-like branches, to 15 m, fls. green: 27719, 27723, 28961*, 30039, 32368. [Elsewhere in Ecuador recorded from lower elevations, < 700 m.]

26. *Lycianthes acutifolia* (Ruiz & Pav.) Bitter
    Sec. for., 1600–2100 m; shrub 2–4 m, corolla white, frs. red: 30201***, 30376, 31690, 31768, 32416, 32793.

27. *Lycianthes* aff. *acutifolia* (Ruiz. & Pav.) Bitter
    Rip. for., Quebrada Santa Rosa, 2000–2050 m; shrub, fls. white: 31244**.

28. *Lycianthes dendriticothrix* Bitter
    1900–2000 m; shrub 1 m, fls. white: Zak & Jaramillo 3239*. [Not in CVPE.]

29. *Lycianthes inaequilatera* (Rusby) Bitter
    Prim. for., 1750–2030 m; shrub 2–4 m, fls. white: 30208**, 30230**, 32858**, 32945**.

30. *Lycianthes radiata* (Sendtn.) Bitter
    Sec. & rip. for., 1400–2300 m; shrub 2 m., fls. white: 30050***, 31752; F 1443.

31. *Markea spruceana* Hunziker
    Rip. for., Río Alambi, 2200–2500 m; ep. shrub to 3 m, corolla purple or greenish: Sparre 15992 [type coll., cited by Knapp, 1998].

32. Nicandra physalodes (L.) Gaertn.
    Disturbed areas, 1950–2000 m; herb c. 0.5 m, frs. green: 31612. "Ambo"

33. *Physalis peruviana* L.
    Cleared areas, 1800 m; shrub 1 m: C 13108. "Uvilla"

34. *Physalis pubescens* L.
    Sec. for., 1250–1400 m; herb with yellow fls.: 28054; Z 153, 187. "Uvilla"

35. *Schultesianthus leucanthus* (Donn. Sm.) Hunz.
    Prim. for., 1800–1900 m; liana with yellow fls.: 29264*.

36. *Solanum acerifolium* Dunal
    Sec. for., 1200–1400 m; shrub 1–1.5 m, fls. yellow: 27232, 27401*, 27918; Z 36; Filskov 3711*.

37. *Solanum aphyodendron* S. Knapp
    Sec. for., 1300–2200 m; shrub or tree 5–10 m, fls. white: 32906***; Zak & Jaramillo 3256***.

38. *Solanum asperolanatum* Ruiz & Pav.
    Sec. for., 1400–2200 m; tree or shrub 3–6 m, fls. white: 27142***, 27239, 28082***; Z 226; Zak & Jaramillo 3234***. "Tambor"

39. *Solanum barbulatum* Zahlbr.
    Prim. for., 1800–2000 m; shrub 4 m, fls. & frs. green: C 7194***.

40. *Solanum brevifolium* Dunal
    Sec. for., 1250–1600 m; herb. vine, corolla white, frs. red-orange: 27004***, 27761, 31007***, 31729. "Veneno de perro", "Tomatillo de monte"

41. *Solanum candidum* Lindl.
    Sec. for. & pastures, 1300 m; shrub 1–1.5 m, fls. white: 27406*. "Berenjena"

42. *Solanum caripense* Dunal
    Prim. for., 2000–2100 m; herb or vine, calyx violet, corolla white: 32853; P 3639***.

43. *Solanum cornifolium* Dunal
    Upper mont. for., Cerro Montecristi, 2500–2700 m; tree 5 m, fls. white: 30598***.

44. *Solanum cucullatum* S. Knapp
    Prim. & sec. for., 1200–2000 m; shrub or tree to 8 m, fls. white, frs. green, 5–6 cm in diam.: 27257*, 27554*, 27922, 28248*, 28698, 31016***, 31253, 31599; C 5931*; Espinoza 729*. [Some specimens were determined as *S. abitaguense* S. Knapp.]

45. *Solanum* cf. *dolosum* S. Knapp
    Prim. for., 1800–2000 m; treelet 5 m: C 7172***.

46. *Solanum furcatum* Dunal
    Sec. for., 1400 m; herb: Filskov 37018.

47. *Solanum juglandifolium* Dunal
    Sec. for., 1200–2500 m; clambering vine, fls. yellow: 27076, 28043, 28187, 28318, 30496, 31776, 32767; C 13051; L 13645; V 12268.

48. *Solanum laevigatum* Dunal
    Upper mont. for., Cerro Montecristi, 2500 m; shrub 0.5 m, corolla white: 30573***.

49. *Solanum lepidotum* Dunal
    Sec. for., 1250 m; tree 7 m, fls. pale green: 29052*; T 204*.

50. *Solanum macrotomum* Bitter
Sec. for., 1900–1950 m; herb 1 m, fls. white: 30110 (det. N. W. Sawyer).

51. *Solanum nigrescens* M. Martens & Galeotti
Sec. for., 1350–2750 m; weedy herb, fls. white: 31218***, 31514; Z 185; Croat 72877***. "Hierba mora"

52. *Solanum nudum* Dunal
Sec. for., 1200–1250 m; shrub 2 m, fls. white: 28309*.

53. *Solanum nutans* Ruiz & Pav.
Sec. for., 1725–2475 m; shrub or tree 4–5 m, fls. white: 31931*; F 1289, 1395, 1420.

54. *Solanum ovalifolium* Dunal
Sec. for., 1200–1550 m; weedy shrub or tree to 6 m, fls. white: 27413***, 29250, 30263***, 31100; Z 13. "Sauco lanudo"

55. *Solanum quitoense* Lam.
Sec. for., 1375 m; shrub, leaves purplish beneath, fls. white: C 6856***; Z 225. [These collections represent the cultivated "Naranjilla", which is widely grown in the area, but escapes to some extent; although presumably native to Ecuador, it may well be introduced in the Nanegal region.]

56. *Solanum rudepannum* Dunal
Sec. for., 1200–1350 m; shrub or tree to 3.5 m, fls. white: 28147. "Tululucha"

57. *Solanum schlechtendalianum* Walp.
Prim. & sec. for., 1400–1750 m; shrub or tree to 6 m, fls. white: 27206, 28756, 30417.

58. *Solanum ternatum* Ruiz & Pav.
Sec. for., 1300–1350 m; creeping vine, leaves pinnatifid: 29086 [specimen barren, determination provisional].

59. *Trianaea nobilis* Planch. & Linden
Prim. & sec. for., 1250–2300 m; liana with pendent yellow-green fls.: 27276*, 28924, 29288*, 31268*, 32942; F 1086(?); Zak & Jaramillo 3229*.

STAPHYLEACEAE (2)
Ref.: Macbride, J. F. 1951. Field Mus. Nat. Hist. Bot. 13(3A,1): 233–235.

1. *Huertea glandulosa* Ruiz & Pav.
Prim. for., 1500–1700 m; tree to 12 m, leaves pinnate, frs. green: 32756; G 73136; Q 35.

2. *Turpinia occidentalis* (Sw.) G. Don
Prim. for., 1500–2300 m; tree 10–18 m, leaves trifoliolate: C 7191 (det. R. Liesner); F 1423, 1468; Q 4. "Cuero de puerco"

STYRACACEAE (1)
Ref.: Gonsoulin, G. J. 1974. Sida 5: 191–258.

1. *Styrax argenteus* C. Presl var. *ramirezii* (Greenm.) Gonsoulin
Prim. for., 1750–2470 m; tree 12–20 m, fls. pink: 30612; F 1239, 1276, 1367, 1497. [Not in CVPE.]

SYMPLOCACEAE (1)
Ref.: Stahl, B. 1991. Fl. Ecuador 43: 1–57.

1. *Symplocos* sp.
Sec. for., 1225 m; white flowers fallen on trail: 31164 [species uncertain because of incomplete material; differing from the relatively few species of *Symplocos* from northern Ecuador, especially at the low elevation, by the large white flowers with many stamens. Possibly an undescribed species.]

THEACEAE (1) (* det. A. L. Weitzman)
Ref.: Kobuski, C. 1941. J. Arnold Arb. 22: 457–496.

1. *Freziera canescens* Humb. & Bonpl.
Prim. & upper mont. for., 1800–2700 m; tree to 15 m, fls. white: 30447, 30602; C 7185*; K 31984. "Guatzhic"

2. *Freziera reticulata* Bonpl.
Loma Pahuamba, 2300 m; tree 10 m: F 1432.

3. *Freziera verrucosa* (Hieron.) Kobuski
Sec. for., 1300–2150 m; shrub or tree to 12 m, fls. white: 27989*, 29081*, 29259; F 1294; T 148*. "Turo aliso"

THEOPHRASTACEAE (1) (* det. B. Stahl)

1. *Clavija* aff. *eggersiana* Mez
Sec. for., 1250–1300 m; unbranched treelet 1–2.5 m: 32741; F 1510; Stahl 4046*.

THYMELAEACEAE (2)
Ref.: Nevling, L. I. 1959. Ann. Missouri Bot. Gard. 46: 257–358.

1. *Daphnopsis* cf. *grandis* Nevling & Barringer
Prim. & sec. for., 1200–1725 m; shrub or tree 3–15 m, fls. pale yellow, frs. white: 31045, 31054, 31144, 31329, 31650, 31689; T 166 (det. K. Barringer). "Sapan"

2. *Schoenobiblus* sp.
Sec. for., 1600 m: T 600 [perhaps confused with *Daphnopsis grandis*].

TILIACEAE (2)
Ref.: Lay, K. K. 1950. Ann. Missouri Bot. Gard. 37: 315–395.

1. *Luehea* sp.
Sec. for., Cerro Santa Lucia, 1600 m; tree 15 m: 30350.

2. *Triumfetta grandiflora* Vahl
Sec. for., 1300–1675 m; shrub or tree to 10 m, corollla yellow: 27718, 28042, 28188, 28700; T 206; Z 40. "Yausa", "Caudillo", "Balsilla"

3. *Triumfetta mollissima* Kunth
1900–2200 m; shrub to 60 cm, corolla pink: Zak & Jaramillo 3255 (det. W. Meijer).

TOVARIACEAE (1)

1. *Tovaria pendula* Ruiz & Pav.
Prim. & sec. for., 1500–2200 m; shrub 1–3 m, inflors. pendulous, fls. greenish: 28083, 30210, 31621, 32394, 32750; P 3616; Zak & Jaramillo 3216.

TROPAEOLACEAE (1) (* det. P. M. Jørgensen)
 Refs.: Sparre, B. 1973. Fl. Ecuador 2: 1–30; Sparre, B. & L. Andersson. 1991.
 Opera Bot. 108: 1–139.

1. *Tropaeolum adpressum* Hughes
 Prim. & sec. for., 1250–1950 m; vine, calyx scarlet with green tip, petals
 purple: 27677*, 27734, 28855, 30329, 31580, 31771; Sparre 19327(?) ("at
 Alambi").

2. *Tropaeolum* cf. *pubescens* Kunth
 Prim. for., 2000–2100 m; vine, calyx orange-red with green tip: 30177, 31535.

3. *Tropaeolum stipulatum* Buchenau
 Upper mont. for., 2450 m; vine, calyx pink with yellow-green tip: 29490.

4. *Tropaeolum* sp. 1
 Sec. for., 1300–1350 m; vine: 29107.

ULMACEAE (2)

1. *Trema micrantha* (L.) Blume
 Sec. for., 1250–2200 m; tree 5–12 m, fls. greenish: 27930, 29258, 29336;
 G 73192; Z 161; Zak & Jaramillo 3223 (det. R. Liesner). "Sapán", "Chale"

2. *Lozanella permollis* Killip & C. V. Morton
 Loma Pahuamba, 2300 m; tree 10 m: F 1449.

URTICACEAE (5)
 Refs.: Killip, E. P. 1939. Contr. U. S. Nat. Heb. 26: 475–530; 1960. Ann.
 Missouri Bot. Gard. 47: 179–198.

1. *Boehmeria aspera* Wedd.
 Sec. for., 1200–1600 m; shrub 1–1.5 m: 27295, 32780.

2. *Boehmeria bullata* Kunth
 Sec. for., 1850–2000 m; shrub 1–4 m: 30454.

3. *Boehmeria caudata* Sw.
 Sec. for., 1250–1700 m; shrub 1–5 m: 27007 (det. R. Liesner), 27243, 32746;
 N 9784; Z 118. "Ortiguilla"

4. *Myriocarpa stipitata* Benth.
   Rip. & sec. for., 1300–2000 m; shrub 2–3 m, spikes pendent, yellowish: 31745, 31851. "Ortiguilla macho"

5. *Phenax hirtus* (Sw.) Wedd.
   Sec. to upper mont. for., 1200–2700 m; herb or shrub to 2 m: 27895, 28310, 30532(?), 32862.

6. *Phenax laevigatus* Wedd.
   Rip. for., 2000 m; shrub 3 m: 31838.

7. *Phenax rugosus* (Poir.) Wedd.
   Prim. for., 1800–2500 m: 31219, 32389; C 13095. "Ashpa ortiga"

8. *Pilea* cf. *antioquiensis* Killip
   Prim. & sec. for., 1300–2025 m; herb to 0.5 m, fls. pink or reddish: 27188, 27485, 27892, 28071, 29042, 29176, 29211; Z 43.

9. *Pilea* cf. *fendleri* Killip
   Prim. for., 2000–2350 m; creeping herb: 29453, 31247. [Not in CVPE.]

10. *Pilea* cf. *flexuosa* Wedd.
    Muddy banks in rip. for., 2000 m; terr., 0.3 m high: 31840.

11. *Pilea* cf. *goudotiana* Wedd.
    Prim. & sec. for., 1650–2100 m; ep. or terr., stems pink, fls. greenish: 29212, 31059, 32860.

12. *Pilea hitchcockii* Killip ?
    Upper mont. for., 2725 m; ep. vine on log: 31978.

13. *Pilea involucrata* (Sims) Urb.
    Sec. for., 1250 m; terr., stems reddish, inflors. yellow-green, flattened against upper leaves in one plane: 32905.

14. *Pilea jamesoniana* Wedd.
    Upper mont. for., 2700–2750 m; trailing herb or epiphytic vine: 30528; H 29524.

15. *Pilea microphylla* (L.) Liebm.
    Banks, sec. upper mont. for., 2300 m: K 31961.

16. *Pilea myriantha* Killip
   Sec. for., 1800–2100 m; stems c. 1 m, leaves aromatic, fls. red: 32751, 32856.

17. *Pilea pubescens* Liebm.
   Sec. for., 1400–1500 m; on mossy banks, fls. greenish: 27204.

18. *Pilea* cf. *puracensis* Killip
   Prim. & sec. for., 1500–1750 m; erect, fls. greenish: 27132. [Not in CVPE.]

19. *Pilea* sp. 1
   Rip. for., 2000 m; shrub 1.5 m, with viny brs.: 31817.

20. *Pilea* sp. 2
   Upper mont. for., 2750–2800 m; erect, stems reddish, leaves thick & dull reddish beneath: 31523.

21. *Pilea* sp. 3
   Sec. for., 1250–1275 m; spreading herb, stems rooting at nodes but becoming erect: 31876.

   [Additional collections of *Pilea* remain unidentified to species.]

22. *Urera baccifera* (L.) Gaudich. ex Wedd.
   Prim. & sec. for., 1300–2000 m; urticant shrub 1–1.5 m, inflor. axes purplish: 27327, 27400, 27952, 30223; C 5949; G 69938; Z 96. "Ortiga"

23. *Urera caracasana* (Jacq.) Griseb.
   Loma Pahuamba, 1750 m; tree 8 m, leaves urticant: F 1517.

VALERIANACEAE (1)
   Ref.: Eriksen, B. 1989. Fl. Ecuador 34: 1–59.

1. *Valeriana chaerophylloides* Sm.
   Sec. for., 1250–1600 m; flaccid herb with white fls.: 27178, 27420, 27931, 32788. [Often listed as *Astrephia chaerophylloides* (Sm.) DC.] "Huanhuan"

2. *Valeriana clematitis* Kunth
   Prim. for., 21 km west of Calacalí, 2500 m: L 13641 (det. B. Erikson).

3. *Valeriana scandens* L.
   Sec. for., El Pacchal, 1300–1500 m; vine with white fls.: 27875, 31886.

VERBENACEAE (7) (* det. R. Liesner)
Ref.: Moldenke, H. N. 1973. Ann. Missouri Bot. Gard. 60: 41–148.

1. *Aegiphila alba* Moldenke
Sec. for., 1200–2000 m; tree 5–15 m, fls. white, fragrant: 28394, 29035, 30519, 32843; N 8642*; Q 81*; T 165*; Z 69; 234. [30519 differs in its more distant lateral veins & conspicuous reticulation, and may not be conspecific]. "Palo blanco", "Jiguerón"

2. *Callicarpa* sp. ?
Sec. for., Cerro Negro, 1500–2000 m; tree 6 m: 28032 [flowers only in bud; perhaps not this genus].

3. *Cornutia microcalycina* Moldenke
Sec. for. along Río Umachaca, 1200–1350 m; tree 12–18 m, corolla purple: 28145, 31169; T 192*; Z 148. [Not in CVPE. Our form best matches var. *pulverulenta* Moldenke (Repert. Sp. Nov. 40: 175. 1936); Ecuadorean plants, including T 192, have been called *C. pyramidata* L., considered by Moldenke to be a Caribbean species; our plants have been collected at higher elevations than species of *Cornutia* elsewhere in Ecuador.]

4. *Duranta repens* L. (*s. lat.*)
Sec. for., east of Tandayapa, 1700 m; shrub 2 m, branches pendent, frs. yellowish: 32763.

5. *Duranta sprucei* Briq.
Prim. for., 1800–2000 m; shrub, corolla turquoise at base, lobes white tinged with purple: C 7174.

6. *Lantana camara* L.
Sec. for., 1200–1300 m; common shrub 1.5 m, fls. orange & red: 27084; Z 44. "Supirrosa"

7. *Stachytarpheta cayennensis* (Rich.) M. Vahl
Sec. for., 1300–1350 m; common herb with blue fls.: 27427; Z 158. "Verbena negra", "Verbena azul"

8. *Verbena litoralis* Kunth
Sec. for., 1200–2000 m; common herb with bluish fls.: 27436, 32846; Z 27; Filskov 37007, 37125 (both det. B. Øllgaard). "Verbena"

VIOLACEAE (1)
Ref.: Smith, L. B., & A. Fernández-Pérez. 1954. Caldasia 6: 83–181.

1. *Viola scandens* Willd. ex Roem. & Schult.
   Prim. for., 2175–2400 m; herb with yellow fls.: 29180, 32811; Rubio 2422.

VITACEAE (2) (* det. J. A. Lombardi)
Ref.: Elias, T. S. 1968. Ann. Missouri Bot. Gard. 55: 81–92; Lombardi, J. A. 1995. Taxon 44: 195–206; 1997. Novon 7: 182–185.

1. *Cissus alata* Jacq. ?
   Upper mont. for., 2750–2800 m; vine with trifoliolate leaves, greenish fls.: 31531 [determination provisional; similar to *C. obliqua*, but stems & leaves distinctly pubescent].

2. *Cissus anisophylla* Lombardi
   Sec. for., 1300–1750 m; vine with coarse, simple leaves, frs. large & brown, often producing pseudofruits due to fungal infection (*Mycosyrinx cissi*): 29118*, 30024*, 30330*, 31728*, 32959*; Z 81, 181. "Escoba de bruja"

3. *Cissus obliqua* Ruiz & Pav.
   Sec. & rip. for., 1900–2400 m; vine with trifoliolate leaves, greenish fls.: 31537*, 31846*, 32448*, 32813. "Mano de sapo"

4. *Cissus verticillata* (L.) Nicolson & C. E. Jarvis
   Sec. for., 1200–1750 m; vine with simple leaves, green fls.: 28276*. [Listed in most manuals as *C. sicyoides* L.] "Bejuco de agua"

5. *Vitis tiliifolia* Humb. & Bonpl. ex Roem. & Schult.
   Sec. for., 1300–1500 m; common liana: 27884. "Bejuco de agua"

MONOCOTS

AGAVACEAE (1)
Ref.: Cerón, C. E. 1994. Hombre y Ambiente 31: 5–38.

1. *Furcraea* cf. *andina* Trel.
   Fields & pastures, 1200–1250 m; plants 10–12 m, fls. greenish-white: 28299, 31000. "Cabuyo blanco"

ALSTROEMERIACEAE (1) (* det. M. Neuendorf)
Refs.: Sodiro, L. 1908. Sert. Fl. Ecuador. 2: 43–57. Neuendorf, M. 1977. Bot.
Notiser 130: 55–60.

1. *Bomarea multiflora* Mirb. ssp. *caldasii* (Kunth) Neuend.
Prim. and upper mont. for. & scrub, 2000–2750 m; perianth orange: 28097,
30153*, 31985; F 1284; H 29544.

2. *Bomarea pardina* Herb.
Prim. & sec. for., 1300–2250 m; vine hanging from trees, sepals rose-pink,
petals white spotted with purple: 28954, 29079, 30472*, 30495, 31142*,
31195, 32837, 32939. [This species with spectacularly showy flowers has
been placed in a new section by Neuendorf (1977).]

3. *Bomarea patocosensis* Herb.
Sec. for., 1800–2000 m; vine with purplish bracts, perianth yellowish, with
purplish spots inside: C 7180 (det. R. E. Gereau, as *Bomarea foliosa* Sodiro).

AMARYLLIDACEAE (1)
Ref.: Meerow, A. W. 1990. Fl. Ecuador 41: 1–52.

1. *Eucrosia dodsonii* Meerow & Dehgan
At edge of banana plantation, Hacienda El Carmen, 1250 m; fls. yellow:
29550.

ARACEAE (8) (* det. T. Croat)
Refs.: Sodiro, L. 1903. Antúrios ecuatorianos. 1903. Engler, A. 1905. Das
Pflanzenreich IV.23B (Heft 21): 1–330. Engler, A., & K. Krause. 1913. Das
Pflanzenreich IV.23Db (Heft 60): 1–143. Croat, T. B. 1983. Ann. Missouri
Bot. Gard. 70: 211–420. Croat, T. B. & N. Lambert. 1986. Aroideana 9: 3–213.

[Presentation of the taxa of Araceae would have been impossible without the
generous assistance of Dr. Thomas Croat (MO). The interpretation of the many
names published by Sodiro in *Anthurium* is difficult, and the synonymy given
here must be regarded as provisional.]

1. *Anthurium argyrostachyum* Sodiro (incl. *A. pandurifolium* Sodiro)
Prim. & sec. for., 1100–1750 m; terr. or hemiep., spathe & spadix green:
27152*, 27338*, 28795*, 30409*; G 73235*; Asplund 17268; Croat 38882,
38902, 38909, 82724; Nanegal Valley, scandent, 3/1900, Sodiro s.n.

2. *Anthurium aristatum* Sodiro
Sec. for., 1175–1600 m; ep. vine, spathe & spadix green or reddish: 27915*, 28748*, 28813*, 29347*, 30161*, 30351*.

3. *Anthurium brachypodum* Sodiro
Prim. & sec. for., 1600–2000 m; terr., to 1 m, spathe green, spadix whitish: 30035*, 30140*.

4. *Anthurium citrifolium* Sodiro
Nanegal Valley; scandent: 1901, Sodiro s.n.; Nanegal, 7/1903, Sodiro s.n. [*A. citrifolium* var. *verruculosum* Sodiro].

5. *Anthurium clathratum* Sodiro
Nanegal; terr.: Sodiro s.n. [known only from the type; status somewhat uncertain].

6. *Anthurium cordiforme* Sodiro
Prim. for., 1550–1700 m; hemiep.: C 39749; G 69922*, 73123*, 73155*.

7. *Anthurium corrugatum* Sodiro (incl. *A. dictyophyllum* Sodiro)
Sec. for., 2100 m; scandent: Smith 1974*; Sodiro s.n.

8. *Anthurium cupulispathum* Croat & J. Rodr.
Sec. for., 1500–1700 m; ep., spathe green, spadix purple: N 9795*.

9. *Anthurium divaricatum* Sodiro
Subtrop. for., Nanegal Valley; scandent: 3/1900, Sodiro s.n. [status somewhat uncertain].

10. *Anthurium dolichostachyum* Sodiro
Prim. for., 1500–2050 m; terr. or ep., spathe green, spadix white or greenish to yellow: 28203*, 30207*; F 1031, 1061; G 69910*, 73217*; C 31718; Croat 50240, 50237A, 72887, 72896, 82704; Hurtado 1427*. "Puchsi"

11. *Anthurium giganteum* Engl.
Sec. for., 1300–1800 m; terr., spadix reddish: 27242*, 27335*, 27997*; C 12426*; G 73148*, 73159*; Croat 38888, 72892; Filskov 37129*.

12. *Anthurium gualeanum* Engl.
Nanegal; terr.: 1900, Sodiro s.n. [Synonym *A. cochliodes* Sodiro; said by Sodiro to be the most robust of all the Ecuadorean species, with leaves to 1.5 × 0.9 m.]

[*Anthurium incomptum* Madison, between Calacalí & Nanegalito, 2175 m, Croat 72886, is just east of our boundary.]

13. *Anthurium incurvatum* Engl.
Nanegal; scandent: Sodiro s.n.

14. *Anthurium jimenae* Croat
Upper mont. for., Cerro Montecristi, 2500 m; ep., petioles over 1 m, spathe white or pinkish, spadix pink: 30581*.

15. *Anthurium laciniosum* Sodiro
Nanegal Valley; scandent: Sodiro s.n.

16. *Anthurium lancea* Sodiro
Prim. & sec. for., 1200–2225 m; mostly terr., sometimes ep. or scandent, lvs. to 1.5 m, spathe & spadix green: 27148*, 27827*, 28197*, 29206, 29414*, 30494*, 31680, 31706; T 200*; Espinoza 725*.

17. *Anthurium longicaudatum* Engl.
Prim. for., 1350–2225 m; terr. & ep., spathe & spadix green to reddish or brown: 27600*, 29137*, 29139*, 29165*, 29431*, 30188*, 32367; C 5898*; Hurtado 1418*.

18. *Anthurium margaricarpum* Sodiro
Sec. for., 1250–1600 m; ep., spathe green to brownish, purplish, or reddish, spadix green: 27481*, 27595*, 27634, 27652*, 28119*, 28127*, 29121*, 29307*, 29634*, 30294*, 30313*, 30404*, 30508*, 31757, 32917.

19. *Anthurium microspadix* Schott (incl. *A. lepturum* Sodiro, *A. nanegalense* Sodiro)
Prim. for., 1250–2175 m; terr., scrambling, or ep., spathe & spadix green: 27687*, 28728*, 29131*, 29203*, 30277*, 30509*, 31214*, 31740, 32411, 32963; F 1390; Croat 82731. [*A. gracilescens* Sodiro & *A. stenophyllum* Sodiro, based on Sodiro collections in the Nanegal Valley, are probable synonyms.]

20. *Anthurium mindense* Sodiro
    Prim. & sec. to upper mont. for., 1200–2600 m; climbing, spathe mostly green, spadix reddish or purple: 27171*, 27233*, 27306*, 27317*, 27393, 27647, 28035*, 28076, 28269, 28763*, 29418, 29448*, 30413*, 30434b*, 30569*, 31681, 31897; C 5900*; L 14039*; N 8650*; Hurtado 1422. "Piquihua"

21. *Anthurium nigrescens* Engl. (incl. *A. malacophyllum* Sodiro)
    Prim. & sec. for., 1800–2000 m; ep., scandent, spathe & spadix green: 30158*, 30180*; Hurtado 1439*.

22. *Anthurium "nonoense"* Croat (ined.)
    Sec. for., 1450 m; ep. spadix green, curved: Croat 38900.

23. *Anthurium obtusum* (Engler) Grayum
    Subtrop. for., Nanegal Valley; ep.: Sodiro s.n.

24. *Anthurium ochreatum* Sodiro (incl. *A. crebrinerve* Sodiro)
    Prim. & rip. for., 1600–2030 m; terr. or hemiep. vine, spathe pale green, spadix green or yellow: 28230*, 31058*, 31073*, 31252*; F 1033; G 73198*; N 8657*; Z 110; Croat 82730.

25. *Anthurium oreophilum* Sodiro
    Sec. for., 2100 m; ep.: Schwertfeger 21132*.

26. *Anthurium orteganum* Sodiro
    Sec. for., 1200–1600 m; terr. & ep., spathe green, spadix reddish or purple: 27487*, 27654*, 27797*, 27798*, 30036*. [31603, from 1920 m, an ep. vine with yellow spadix, may not belong here]. [Omitted from CVPE.]

27. *Anthurium ovatifolium* Engler (incl. *A. elatius* Sodiro, *A. miconiifolium* Sodiro, *A. pachyphyllum* Sodiro, *A. quitense* Sodiro, *A. soederstroemii* Sodiro)
    Prim. & sec. for., 1200–2250 m; terr. & ep., spathe greenish, spadix brown: 27645*, 27998*, 28793*, 29231, 29429*, 30049*, 30160*, 30426*, 31900, 32427; Croat 72871, 82720; Espinoza 726*; Nanegal Valley, 3/1902, Sodiro s.n. [as *A. miconiifolium*]; from Nanegal Valley near Auca, Sodiro s.n. [as *A. quitense*]; Nanegal, Sodiro s.n. [as *A. soederstroemii*]. "Plato quiro"

28. *Anthurium pallatangense* Engl.
    Prim. for., 1900–2225 m; terr., spathe & spadix green: 29413*, 30189*; Luteyn & Lebrón-Luteyn 6523.

29. *Anthurium patulum* Sodiro
    Nanegal; scandent: 1900, Sodiro s.n.

30. *Anthurium pulverulentum* Sodiro
    Prim. & sec. for., 1400–2550 m; terr. or ep., spathe green, spadix green or
    brownish: 28928*, 29415*, 30068*, 30303*, 30380*, 32373; G 69948*;
    Croat 50229, 50236, 72890.

31. *Anthurium rigidifolium* Engl. (incl. *A. lunatum* Sodiro)
    Nanegal; scandent: Sodiro s.n.

32. *Anthurium sarmentosum* Engl.
    Sec. for., 1250–1500 m; terr. & ep. vine, spathe green, spadix dark red or
    purplish: 27034*, 27646*, 27900*, 31568.

33. *Anthurium scandens* (Aubl.) Engl. ssp. *scandens*
    Sec. for., 1300–2000 m; ep. vine, spathe green, spadix greenish to reddish:
    27584, 27813*, 28117*, 30410*; Hurtado 1438*.

34. *Anthurium striatipes* Sodiro (incl. *A. subtrigonum* Engl.)
    Prim. & sec. to upper mont. for., 1175–2500 m; terr. & ep., petioles c. 1 m
    long, spathe green, spadix yellow or reddish: 27147*, 27492*, 27817*,
    27851*, 27890*, 27909*, 27945*, 28780*, 28807*, 29120*, 29302*, 29477*,
    29980*, 30215*, 30317*, 30422*, 30565*; C 39720, 39735; N 8651*; Harling
    & Andersson 11584; Hurtado 1421*; Nanegal, Sodiro s.n.

35. *Anthurium subcoerulescens* Engl.
    Nanegal; scandent: Sodiro 42.

36. *Anthurium tenuifolium* Engl.
    Prim. & sec. for., 1300–2000 m; usually ep., sometimes climbing, spathe
    mostly green, spadix green or creamy: 27815, 27824, 28324, 29454, 30052*,
    30191*, 30434a*; Q 120.

37. *Anthurium tremulum* Sodiro (incl. *A. buglossum* Sodiro, *A. vexillare* Sodiro)
    Prim. & sec. for., 1500–2100 m; terr., climbing, spathe green, spadix brownish:
    27150*, 27319*, 27562*, 28007*, 28075*, 28927*, 30146*; Nanegal Valley,
    May 1901, Sodiro s.n. [type coll.]; Dec. 1899 & June 1900, Sodiro s.n.;
    4/1900, Sodiro s.n. [syntype coll. of *A. vexillare*].

38. *Anthurium truncicola* Engl.
   Prim. & sec. for., 1200–1750 m; ep. or climbing, spathe green, spadix creamy: 27994*, 28176*, 28543*, 28730*; Croat 38910; Nanegal, Sodiro s.n.; Nanegal, 3/1900, Sodiro s.n. [type coll. of *A. platylobum* Sodiro]; Nanegal Valley, Sodiro s.n. [type coll. of *A. subdeltoideum* Engl.].

39. *Anthurium umbraculum* Sodiro
   Prim. & sec. for., 2175–2550 m; terr., subscandent, lf. blades & petioles each c. 1 m. or more, spathe white or green, spadix green: F 1302, 1371; Croat 50228, 72888; Nanegal, Sodiro s.n.; Sparre 16797*.

40. *Anthurium versicolor* Sodiro (incl. *A. hylaeum* Sodiro)
   Prim. & sec. for., 1250–2050 m; climbing, spathe green, spadix green or white: 27039*, 27040*, 27552*, 27666*, 27944*, 31336*; Croat 27894, 38881, 38908, 50234; Holm-Nielsen 24434*; Hurtado 1415*.

41. *Dieffenbachia daguensis* Engl.
   Clearings in sec. for., 1200–1950 m; terr., spathe green below, yellowish above, sometimes reddish within, spadix creamy, with slight acrid fragrance: 28301, 28541*.

42. *Monstera adansonii* Schott
   Sec. for., 1200–1300 m; scandent: 28162*, 31343, 32912.

43. *Monstera lechleriana* Schott
   Sec. for., 1250–1630 m; ep.: 27321*, 28962*; G 69906*.

44. *Philodendron acuminatissimum* Engl.
   Prim. & sec. for., 1200–1750 m; climbing, spathe greenish to reddish or purplish below, whitish above, spadix white: 27559*, 27948*, 27999*, 28751*, 29282*, 31551, 31731; V 12280*.

45. *Philodendron dodsonii* Croat & Grayum
   Sec. for., 1250–1650 m; ep. or climber, spathe reddish below, white above, spadix white: 27500*, 28781*, 30069*; Croat 38895.

46. *Philodendron "fibrosum"* Sodiro ex Croat (ined.)
   Prim. & sec. for., 1300–1950 m; hemiep., spathe greenish outside, maroon inside, spadix maroon: 27599*, 28074*, 31616*; G 73146*; Asplund 17247; Croat 82732. [Not in CVPE.]

47. *Philodendron gualeanum* Engl.
   Nanegal & Gualea: Apr. 1900, Sodiro 13.

48. *Philodendron inaequilaterum* Liebm.
   Roadcut in sec. for., El Carmen to Marianitas, 1200–1250 m; terr.: 28148*.

49. *Philodendron musifolium* Engl.
   Sec. for., 1700 m; ep.: 30372*; Croat 82736.

50. *Philodendron nanegalense* Engl.
   Nanegal Valley; scandent: Sodiro 12f. [Known only from the type].

51. *Philodendron oligospermum* Engl.
   Prim. & sec. for., 1150–2000 m; climbing vine, mostly ep., spathe white or pink, sometimes green at base, spadix white: 27320*, 27412*, 27560*, 27663*, 27773*, 27995*, 28011*, 28828*, 29125*, 29343*, 29346*, 30046*, 30176, 30405*, 31929, 32410; C 5899*; Croat 72891*; Schwerdtfeger 21115*; Nanegal, 1899, Sodiro.

52. *Philodendron rhodoaxis* G. S. Bunting ssp. *lewisii* Croat & Grayum
   Sec. for., 1250–1400 m; ep.: 30332, 31545 (both det. M. H. Grayum).

   [*Philodendron senatocarpium* Madison, at 2175 m, Croat 72885, is just east of our boundary.]

53. *Philodendron sodiroanum* Engl.
   Nanegal: Aug. 1874, Sodiro 47.   [Not in CVPE, possibly a synonym of *P. tenue* K. Koch.]

54. *Philodendron subhastatum* Engl. & K. Krause
   Prim. & sec. for., 1150–1750 m; climbing vine, terr. or ep., spathe reddish, spadix white: 27044*, 27853*, 27996*, 28396*, 28836*; between Nanegal & Gualea, Sodiro 12g.

55. *Philodendron sulcatum* K. Krause
   Rip. & sec. for., 1250–1725 m; climbing ep. Vine, spathe dark green, spadix green and yellow: 27667, 31887 (both det. M. H. Grayum); Croat 72895.

56. *Philodendron validinervium* Engl.
   Nanegal & Gualea: Sodiro 11 [type collection].

57. *Philodendron verrucosum* L. Mathieu ex Schott
Prim. & sec. for., 1400–1800 m; common climber or ep., spathe reddish inside, spadix white: 27177*, 27748; Z 117.

58. *Philodendron* sp. 1
Sec. for., Cerro Santa Lucia, 1400 m; ep.: 30338. [Possibly a form of *P. dodsonii.*]

59. *Philodendron* sp. 2
Prim. for., 1900–2300 m; ep., scandent, petioles to 1.3 m, leaf blade 1 m, spathe dark red, spadix creamy, with peppery odor: 29228 [28006 and 28922 may also belong here.]

60. *Rhodospatha densinervia* Engl. & K. Krause
Prim. for., 1550–1800 m; climber: 28198*.

61. *Rhodospatha statutii* Sodiro
In "silv. subtrop. vall. Nanegal et Gualea", Sodiro.

62. *Spathiphyllum grandifolium* Engl.
Sec. for., 1175 m; terr., spathe & spadix green: 28808*.

63. *Stenospermation densiovulatum* Engl. (incl. *S. porteri* Sodiro)
Prim., rip. & sec. for., 1250–2000 m; scandent, frs. green: 28196*, 31033*, 31155*; G 73203*.

64. *Stenospermation longifolium* Engl.
Prim. for., 1975–2350 m; terr., spathe green, spadix white: 28712*, 30203*; Croat 50235; Nanegal/Gualea, 1900, Sodiro s.n.

65. *Stenospermation longipetiolatum* Engl.
Sec. for., 1600–1700 m; terr., spathe & spadix whitish: 30433*, 32953; "inter Nanegal et Gualea", Sodiro.

66. *Stenospermation mathewsii* Schott var. *stipitatum* Engl.
Sec. for., Nanegal & vicinity, c. 900–1600 m: Radrigan 17443*; Sodiro.

67. *Stenospermation sparrei* Croat
Upper mont. for., 2350–2700 m; terr., spathe green or brown, spadix white or yellow: 29452*, 29473*, 30562*.

68. *Xanthosoma daguense* Engl.
Prim. & sec. for., 1500–2100 m; terr., spathe green below, yellowish above, spadix creamy: 30488*, 31618; Croat 82709. [Not in CVPE.]

69. *Xanthosoma undipes* (K. Koch & C. D. Bouché) K. Koch
Prim. & sec. for., 1600–2000 m; common giant herb to 3 m, leaves to 2 m, spathe & spadix whitish: 27241*, 27822*, 30218*, 31622. [Commonly listed as *X. sagittifolium* (L.) Schott.] "Camacho"

ARECACEAE (9) (* det. H. Balslev; ** det. B. Bergmann; *** det. F. Borchsenius)
Refs.: Balslev. H. 1990. AAU Rep. 25: 23–26. Balslev, H., & A. Barfod. 1987. Opera Bot. 92: 17–35. Svenning, J. C. & H. Balslev. 1998. Principes 42: 218–226.

1. *Aiphanes chiribogensis* Borchs. & Balslev
Prim. for., slopes above Río Pichán, 1500–2000 m; trunk 1–3 m, fronds 1 m, inflors. erect, purplish: 30169***; F 1178*.

2. *Aiphanes erinacea* (H. Karst.) H. Wendl.
Prim. & sec. for., 1300–1800 m; trunk 2–8 m, inflor. axes purplish, fls. white: 27113, 27262*; C 12415*; G 69925*; N 9794***.

3. *Bactris setulosa* H. Karst.
Between Nanegalito & Nanegal, and Maquipucuna, 1200–1400 m; multistemmed, to 10 m: Balslev [sight observation]. "Chonta fina"

4. *Ceroxylon alpinum* Bonpl. ex DC. ssp. *ecuadorensis* Galeano
Between Nanegalito & Nanegal, and Maquipucuna: Balslev [sight observation]. "Ramos", "Palma de cera", "Palma de ramo"

5. *Chamaedorea linearis* (Ruiz & Pav.) Mart.
Sec. for., less commonly in prim. for., 1300–1800 m; trunk 3–10 m, fronds 1.5–2 m, fls. creamy white, frs. orange: 27499*, 27501*, 31763, 31896; C 13072. "Palmiche", "Chonte verde", "Coco"

6. *Chamaedorea pinnatifrons* (Jacq.) Oerst.
Prim. & sec. for., 1250–2000 m; trunk 2–10 m, frs. orange: 27675**, 28788*, 29285*, 30070***, 30165, 31606, 31692; C 5950*, 39722; F 1022, 1060, 1522*; N 9797*. "Molinillo"

7. *Geonoma undata* Klotzsch
   Prim. & sec. for., 1300–1700 m; fronds 2–2.5 m, inflor. axes orange: 27391; Z 88. "Corozo"

8. *Oenocarpus bataua* Mart.
   Río Alambi between Nanegalito & Nanegal, 1200 m; over 20 m tall: Svenning & Balslev [sight observation].

9. *Phytelephas aequitorialis* Spruce
   Between Nanegalito & Nanegal: Balslev [sight observation]. "Tagua"

10. *Prestoea acuminata* (Willd.) H. E. Moore var. *acuminata*
    Prim. & sec. for., 1250–2000 m; trunk 10–15 m, inflor. axes brownish, fls. pink, frs. blackish: 27476*, 28774*; F 1078, 1104*, 1184*, 1551; C 39767; G 69907, 73149*. "Pambil", "Palmito"

11. *Socratea rostrata* Burrett
    Between Nanegalito & Nanegal, and Maquipucuna, 1100–1300 m: Balslev [sight observation].

BROMELIACEAE (6) (* det. H. Luther; ** det. M. A. Spencer; *** det. M. Manzanares)
Refs.: Gilmartin, A. J. 1972. The Bromeliaceae of Ecuador (Phaner. Monogr. 4). Luther, H. 1989. Phytologia 67: 312–330.

1. *Catopsis sessiliflora* (Ruiz & Pav.) Mez
   Sec. for., 1250–1400 m; ep., fls. white: 27398, 27541, 27543.

2. *Guzmania* cf. *amplectens* L. B. Sm.
   Sec. for., 1600–2000 m; ep., bracts red, sometimes with greenish tips: 30383*, 32961.

3. *Guzmania bracteosa* (André) André ex Mez
   Sec. for., 2100 m; ep., bracts purplish, fls. yellow: Manzanares 3070.

4. *Guzmania* cf. *devansayana* E. Morren
   Prim. for., 1750 m; ep.: 27767. "Huicundo"

5. *Guzmania fuscispica* Mez & Sodiro
   Sec. for., 1200–1250 m; ep., bracts pinkish: 31158*; Filskov 37134*.

6. *Guzmania gloriosa* (André) André ex Mez
   Prim. & upper mont. for., 2200–2750 m; ep., bracts green with red or orange tips, fls. yellow: 29154, 29501***.

7. *Guzmania jaramilloi* H. Luther
   Prim. & sec. for., 1200–2000 m; ep., bracts red, fls. yellow: 28340*, 31057*; C 5935*; N 8652*; T 606*; Hurtado 1410*.

8. *Guzmania lehmanniana* (Wittm.) Mez
   Prim. & sec. for., 1200–1650 m; ep., inflors. 0.5–1 m, bracts reddish-green (tertiary red), fls. white: 31065*; Espinoza 730*.

9. *Guzmania melinonis* Regel
   Prim. & sec. for., 1500–2000 m; ep., bracts red, fls. yellow: 27558, 27695**, 28049; F 1116; Espinoza 723***.

10. *Guzmania monostachia* (L.) Rusby ex Mez
    Sec. for., 1200–1250 m; ep., bracts green with red tips, fls. white: 29978*.

11. *Guzmania musaica* (Linden & André) Mez var. *concolor* L. B. Sm.
    Sec. for., 1300 m; ep.: 27713**.

12. *Guzmania patula* Mez & Wercklé
    Sec. for., 1200–1575 m; ep., bracts yellow-green, fls. whitish: 27540, 27542, 27711**, 29272, 31121.

13. *Guzmania rhonhofiana* Harms
    Sec. for., 1300–1325 m; terr., inflor. axes red, bracts yellow: 31017*.

14. *Guzmania teuscheri* L. B. Sm.
    Prim. for., 1675 m; terr., bracts filled with mucilage: 30365*.

15. *Guzmania variegata* L. B. Sm.
    Prim. for., 1700–2300 m; ep., inflors. 1 m, bracts red: 27112**, 27551**, 29185, 29197, 31677; F 1292***, 1415***; Manzanares 1201. [29494, at 2700–2750 m on Montecristi, may belong here.]

16. *Guzmania wittmackii* (André) André ex Mez
    Sec. for., 1250 m; ep., bracts magenta-tipped (sec. white): 31345*.

17. *Guzmania xanthobractea* Gilmartin
    Sec. for., 1300–1700 m; ep., bracts red (sometimes with green tips), secondary bracts & fls. yellow: 27475**, 31064*, 31117* [atypical], 31919; N 9799*.

18. *Mezobromelia lymansmithii* Rauh & Barthlott
    Banks, east of Nanegalito, 1600–2200 m; terr. or ep., inflors. to 2 m, bracts & fls. yellow: F 1485***; Dodson et al. 6981***.

19. *Mezobromelia pleiosticha* (Griseb.) Utley & H. Luther
    Prim. for., 1250–2100 m; terr. or ep., bracts red with greenish tip, filled with mucilage; fls. white: 30387*, 30503*, 30513*, 31138*.

20. *Pitcairnia* cf. *bakeri* (André) André ex Mez
    Prim. for., 2030 m; terr., 1.3 m: 30209.

21. *Pitcairnia commixta* L. B. Sm.
    Sec. for., 1600 m; terr., inflors. c. 1 m, frs. reddish: 31125*.

22. *Pitcairnia fusca* H. Luther
    Upper mont. for., Cerro Montecristi, 2750 m; terr., 1 m, fls. white or pale yellow: 30579; H 29549*.

23. *Pitcairnia lehmannii* Baker
    Banks in sec. for., 1200–1400 m; terr., leaves nearly 2 m, inflors. 1.5 m, fls. scarlet: 29265, 32382.

24. *Pitcairnia nigra* (Carrière) André
    Prim. & sec. for., 1500–2100 m; ep., bracts reddish, fls. yellow: 27498**, 28739***, 28882, 30029, 30345, 31901; F 1538.

25. *Pitcairnia sceptrigera* Mez
    Sec. for., Marianitas to Montecristi, 1200–2700 m; terr., leaves & spikes c. 2 m, fls. yellow: 28316; K 31992.

26. *Pitcairnia sodiroi* Mez
    Prim. & upper mont. for., 1750–2550 m; terr. or ep., fls. orange: 28659, 28879, 28888, 29191, 29424, 30576*, 32417; C 7176*; F 1271.

27. *Racinaea elegans* (L. B. Sm.) M. A. Spencer & L. B. Sm.
    Upper mont. for., 2500–2750 m; basket ep. to 1.5 m across, inflors. 1 m, fls. yellow: 29478, 30546*, 31509.

28. *Racinaea ropalocarpa* (André) M. A. Spencer & L. B. Sm.
Sec. for., 1900 m; ep.: 31644.

29. *Racinaea tandapiana* (H. Luther) M. A. Spencer & L. B. Sm.
Loma Pahuamba, 2470 m; ep., bracts brownish, fls. green: F 1376***.

30. *Racinaea tetrantha* (Ruiz & Pav.) L. B. Sm. & M. A. Spencer var. *miniata* (André) L. B. Sm. & M. A. Spencer
Upper mont. for., Cerro Montecristi, 2750 m; ep., bracts orange, calyx yellow: 30522*.

31. *Tillandsia* cf. *buseri* Mez
Sec. for., Cerro Negro, 1800–1850 m; terr., bracts red, secondary bracts yellow: 31943.

32. *Tillandsia complanata* Benth.
Prim. for., 1950–2150 m; ep.: 30102*, 30175*; C 5936*; F 1212. "Huicundo"

33. *Tillandsia pretiosa* Mez
Sec. for., 1500–1650 m; ep., bracts pink or orangish, fls. purple: 30063*, 31902.

34. *Tillandsia truncata* L. B. Sm.
Prim. for., 1950–2200 m; ep., bracts red; fls white or green: 30193*; C 5937*; F 1194, 1477**.

35. *Tillandsia venusta* Mez & Wercklé
Sec. for., 1300 m; ep.: 28181.

CANNACEAE (1)
Ref.: Maas, P. J. M., & H. Maas. 1988. Fl. Ecuador 32: 1–9.

1. *Canna jaegeriana* Urb.
Sec. for., 1300–2000 m; herb to 3 m, fls. yellow-orange, frs. yellowish-green with purple indumentum: 27724, 31188, 31849.

COMMELINACEAE (5) (* det. R. Faden)
Ref.: Macbride, J. F. 1936. Publ. Field Mus. Nat. Hist., Bot. 13(1): 592–608.

1. *Callisia gracilis* (Kunth) D. R. Hunt
Sec. for., clearings, 1250–1500 m; fls. white: 27657, 27946*; Z 70, 179.

2. *Commelina diffusa* Burm. f.
    Sec. for., 1500 m; spreading herb, fls. blue: Z 56. "Suelda"

3. *Commelina obliqua* Vahl
    Sec. for., 1150–1500 m; fls. blue: 27416, 27887*, 28128*, 28803*; Z 62, 142.
    "Cacharillo"

4. *Elasis hirsuta* (Kunth) D. R. Hunt
    Upper mont. sec. for., 2750 m; common, fls. pink: 31512; K 31976.

5. *Tradescantia zanonia* (L.) Sw.
    Sec. for., 1200–2000 m; common, to 1.5 m, fls. white: 27308, 27889, 29329,
    30290, 32799, 32930; Hurtado 1444 (det. P.J.M. Maas). [Commonly treated
    as *Campelia zanonia* (L.) Kunth.]

6. *Tripogandra serrulata* (Vahl) Handlos
    Sec. for., 1250–1725 m; weed with white fls.: 27191*, 27891*, 27924*, 31927,
    32931; Z 204.

COSTACEAE (1)
    Ref.: Maas, P. J. M. 1976. Fl. Ecuador 6: 1–49.

1. *Costus guanaiensis* Rusby
    Sec. for., 1200–1300 m; giant herb to 4 m, bracts red, fls. yellowish-white:
    27624, 31272.

2. *Costus laevis* Ruiz & Pav.
    Prim. & sec. for., 1200–1800 m; bracts red, fls. striped red & yellow: 27437,
    28213A, 28313, 29996. "Caña agria"

3. *Costus pulverulentus* C. Presl
    Sec. for., 1200–1400 m; 1–5 m, bracts red: 27285, 27517, 27588, 28289,
    28837.

CYCLANTHACEAE (4) (* det. R. Erikson)
    Ref.: Harling, G. 1973. Fl. Ecuador 1: 1–48.

1. *Asplundia stenophylla* (Standl.) Harling
    Prim. for., 1550–1900 m; hemiep. climber, inflors. green: C 39728; F 1025;
    G 73154 (det. B. Hammel). [Not in CVPE.]

2. *Asplundia* sp. 1 (subg. *Choanopsis*)
   Prim. for., 1700–1800 m; ep., inflors. green: 28210*.

3. *Asplundia* sp. 2 (subg. *Asplundia*)
   Sec. for., 1300–1400 m; terr., uncommon: 29303*.

4. *Asplundia* sp. 3 (subg. *Asplundia*)
   Sec. for., 1700 m; ep.: 30071*.

   [Additional specimens of *Asplundia* remain unidentified.]

5. *Cyclanthus bipartitus* Poit.
   Sec. for., 1300–1500 m; leaves 1.5–2 m: 27899, 31903; Z 103.

6. *Ludovia bierhorstii* G. Wilder
   Sec. for., 1250–1500 m; hemiep. climber, bracts & spikes yellow: 28273*,
   31552.

7. *Sphaeradenia hamata* Harling
   Prim. for., 1550–1800 m; ep. or terr., sometimes climbing, fls. & frs. white:
   27111*, 27135*, 27477*, 27549*, 27803*, 28214*, 28764*, 31676, 31891;
   G 73151, 73251 (both det. M. Merello).

8. *Sphaeradenia horrida* (Harling) Harling
   Prim. for., 1400–2750 m; ep. or terr.: 28955*, 29202*, 29447*, 30521, 31053,
   31317; F 1200, 1303; K 31977.

CYPERACEAE (9) (* det. G. Davidse; ** det. T. Reznicek; *** det. K.
Camelbeke)
Refs.: Clarke, C. B. 1905. Bot. Jahrb. 34, Beibl. 78: 5. Tucker, G. C. 1983.
Syst. Bot. Monogr. 2: 1–85 [*Cyperus*]. 1987. J. Arnold Arb. 68: 361–445.

1. *Carex jamesonii* Boott
   Prim. for., along stream, Quebrada Santa Rosa, 2000 m: 31236**, 32420.

2. *Carex polystachya* Sw. ex Wahlenb.
   Sec. for., 1300–2500 m: 31149**; K 31805; L 13662.

3. *Carex* sp. 1
   Banks, 1300 m: K 32118.

4. *Cyperus aggregatus* (Willd.) Endl.
   Sec. for., 1200–2000 m: 28280, 29067, 32393; Z 163.

5. *Cyperus chalaranthus* J. Presl & C. Presl
   Sec. for., scrub, roadsides, 1200–1400 m: 27418, 28108***, 28279, 29066; Z 11; Filskov 37087.

6. *Cyperus hermaphroditus* (Jacq.) Standl.
   Pastures & along streams, 1400–1920 m: 30114, 32398; Z 222.

7. *Cyperus laxus* Lam.
   Fields, 1200–1400 m: Filskov 37089 (det. by S. Laegaard as *C. diffusus* Vahl).

8. *Cyperus luzulae* (L.) Rottb. ex Retz.
   Fields & roadsides, 1200–1350 m: 27161, 27424; Z 171. "Zacate estrella", "Cabezona"

9. *Cyperus mutisii* (Kunth) Griseb.
   Fields, 1250–1300 m: 28151***.

10. *Cyperus odoratus* L.
    Ditches, 1600 m: 32795; Z 57.

11. *Cyperus prolixus* Kunth
    Moist fields between Nanegal & Nanegalito, 1200–1300 m; stems to 1.5 m: 28899*.

12. *Cyperus simplex* Kunth
    Ditches, 1700–1900 m: 31637; Laegaard 53111.

13. *Eleocharis elegans* (Kunth) Roem. & Schult.
    Sec. for., 1250–2100 m: 27162, 27165, 30489, 31349; C 13125.

14. *Eleocharis geniculata* (L.) Roem. & Schult.
    Wet banks, 1250 m; diminutive annual: 31569.

15. *Eleocharis maculosa* (Vahl) Roem. & Schult.
    Sec. for., 1250–1300 m: 27934***.

16. *Eleocharis retroflexa* (Poir.) Urb.
    Fields & sec. for., 1250–1300 m: 28131 (det. J. Bruhl).

17. *Fimbristylis dichotoma* (L.) Vahl
   Roadsides, 1200–1250 m: 27164; Z 157.

18. *Kyllinga pumila* Michx.
   Sec. for., 1250–1350 m: 27381, 27425; Z 58.

19. *Pycreus bipartitus* (Torr.) C. B. Clarke
   Fields in sec. for., 1250 m: 28140***.

20. *Pycreus niger* (Riuz & Pav.) Cufod.
   Bed of Río Pichán, 1900 m: 32399.

21. *Rhynchospora polyphylla* Vahl
   Sec. for., 1250–2750 m: 27441**, 27660, 28719**, 30356*, 31502.

22. *Rhynchospora polystachys* (Turrill) H. Pfeiffer
   Sec. for., 1250–1300 m: 27941*.

23. *Rhynchospora puberula* (Boeck.) L. B. Sm.
   Sec. for., 1250–1850 m: 27936**, 28852**, 30051. [=*Pleurostachys puberula* Boeck.]

25. *Rhynchospora radicans* (Schltdl. & Cham.) H. Pfeiffer
   Sec. for., 1300–1400 m: 27155.

26. *Rhynchospora ruiziana* Boeck.
   Rip. for., 2000 m: 31235**, 32423; Sodiro 43e.

27. *Rhynchospora vulcani* Boeck.
   Sec. for., Cerro Negro, 1500–2000 m: 28029**.

28. *Rhynchospora watsonii* (Britton) Davidse
   Banks, 1250–1550 m: 31781, 32355.

29. *Rhynchospora* sp. 1.
   Prim. for., 2750 m; inflors. to 1.5 m: H 29542.

30. *Scleria microcarpa* Nees ex Kunth
   Nanegal Valley: Sodiro 199/55. [*Scleria sororia* Kunth, cited by Clarke on the basis of Sodiro 199/53 from Nanegal Valley & Río Pilatón, requires confirmation.]

31. *Uncinia hamata* (Sw.) Urb.
    Prim. for., 1800–2215 m: 29129, 29411; C 13068. "Cortadira", "Pega-pega"

DIOSCOREACEAE (1)
Ref.: Ayala F., F. 1981. Ann. Missouri Bot. Gard. 68: 125–131.

1. *Dioscorea* cf. *coriacea* Humb. & Bonpl. ex Willd.
    Sec. for., 1150 m; vine with reddish fls.: 29075.

2. *Dioscorea pilosiuscula* Bertero ex Spreng.
    Sec. for., 1900–2200 m; vine with dull purple fls.: Zak & Jaramillo 3253 (det.
    O. Tellez).

3. *Dioscorea polygonoides* Humb. & Bonpl. ex Willd.
    Sec. for., 1800–2000 m; vine, fls. greenish: 31538, 31942.

4. *Dioscorea* cf. *syringifolia* Kunth & Schomb. ex Schomb.
    Sec. for., 1300–1500 m; vine with purplish fls.: 27910. [Not in CVPE.]

5. *Dioscorea* sp. 1
    Prim. for., 2000–2375 m; vine: 29462.

HAEMODORACEAE (1)

1. *Xiphidium caeruleum* Aubl.
    Sec. for., 1200–1250 m; fls. whitish, frs. orange: 27373, 28165.

HELICONIACEAE (1) (* det. W. J. Kress)
Refs.: Andersson, L. 1985. Fl. Ecuador 22: 1–86. 1985, 1992. Opera Bot. 82:
1–123; 111: 1–98.

1. *Heliconia burleana* Abalo & G. L. Morales
    Prim. for., 1700–2000 m; stem 1.5 m, leaves to 2.5 m, inflors. erect, bracts red,
    ovary greenish or yellow, perianth yellow or green: 27561*, 27854*, 30219*,
    30373, 30378, 31223*; Morales & Abalo 294 [type collection] from
    Tandayapa, 2050 m.

2. *Heliconia griggsiana* L. B. Sm.
    Sec. for., 1200–1300 m; to 8 m, leaves 1.5 m, inflors. pendent, bracts red, fls.
    yellow, frs. orange: 27623*, 29046*; C 13071; Filskov 37135 (det. L.
    Andersson).

3. *Heliconia impudica* Abalo & G. L. Morales
   Sec. for., 1600–2000 m; to 2 m high, inflors. erect, bracts red or red & yellow, fls. yellow or orange: 27102*, 27622*, 28063*, 28200*; C 5934*. "Platanillo"

4. **Heliconia pardoi** Abalo & G. L. Morales
   Rip. sec. for., 1400 m; 3 m high, inflors. erect, fls. yellow, frs. green: 27273*.

5. *Heliconia sclerotricha* Abalo & G. L. Morales
   Sec. for., 1250–1600 m; to 2 m high, inflors. pendent, bracts red, fls. yellow: 27375*, 27466.

6. *Heliconia virginalis* Abalo & G. L. Morales
   Prim. & sec. for., 1250–1700 m; to 3.5 m, leaves 1 m, inflors. erect, bracts red, fls. white, frs. yellow turning blue: 27173*, 27414*, 27488*, 27564*, 27742*, 27788*, 29300*, 30272, 30295, 31741.

HYPOXIDACEAE (1)

1. *Hypoxis decumbens* L.
   Sec. for., 1200–1300 m; fls. yellow: 27386, 27799; Z 194.

IRIDACEAE (1)

1. *Sisyrinchium micranthum* Cav.
   Open areas, 1250–1350 m; fls. lavender: 27371, 27410, 28110.

JUNCACEAE (1)
Ref.: Balslev, H. 1979. Fl. Ecuador 11: 1–45.

1. *Juncus tenuis* Willd.
   Open areas, 1300–2300 m: 28030, 31636, 32820; Acosta Solís 19150.

MARANTACEAE (3) (* det. H. Kennedy)
Ref.: Kennedy, H., L. Andersson & M. Hagberg. 1988. Fl. Ecuador 32: 11–191.

1. *Calathea ischnosiphonoides* H. Kenn.
   Prim. for., 1500–1800 m; to 3 m, bracts brownish, fls. white or pink: 27331, 28204*, 31700; N 9801*.

2. *Calathea plurispicata* H. Kenn.
  Sec. for., 1300–1400 m; leaves to 4 m, inflors. to 2.5 m: 29325*.

3. *Calathea selbyana* H. Kenn.
  Sec. for., 1350–1400 m; plants 1.5 m, bracts green: 30274*.

4. *Pleiostachya pruinosa* (Regel) K. Schum.
  Sec. for., 1100 m; leaves 1.5 m, fls. blue: 29266*.

5. *Stromanthe stromanthoides* (J. F. Macbr.) L. Andersson
  Sec. for., 1200–1600 m; leaves to 2 m, bracts orange, fls. whitish: 27491*,
  29327*, 31021, 31091.

[Musaceae, Musa acuminata Colla, at 1200 m; cult.: Filskov 37136 (det. J. Kress).]

ORCHIDACEAE (66) (* det. C. Dodson; ** det. J. Beckner)
  Refs.: Garay, L. 1978. Fl. Ecuador 9: 1–304. Dodson, C. H. & P. M. Dodson.
  1978–1984. Icones Plantarum Tropicarum, Ecuador vols. 1–5, 10; series 2,
  vols. 5, 6, 1989. Luer, C. A. 1986. Missouri Bot. Gard. Syst. Monogr. 20: 1–
  109. C. H. Dodson & R. Escobar R., 1996. Orquídeas Nativas del Ecuador,
  vol. I.

  [A large number of orchids have been collected by Dodson, Luer, and others
  along the road between Nono & Tandayapa, but most of these lie outside our
  boundaries; those within approximately 1′ are listed in brackets.]

1. *Ada glumacea* (Lindl.) N. Williams
  Prim. for., Tandayapa, 1600 m; ep.: Dodson 18794.

2. *Ada ocanensis* (Lindl.) N. Williams
  Prim. for., Tandayapa, 2000 m; ep.: Dodson 1097; Luer 4711*. [32921, at
  1300–1350 m, has the habit of *Ada ocanensis*, but the flower color (green with
  white lip) of *Ada andreettae* Dodson.]

  [*Altensteinia fimbriata* Kunth: Calacalí to Nanegalito, 2400 m, K 1015.]

3. *Barbosella cucullata* (Lindl.) Schltr.
  Nono–Nanegal, km 17, 2160 m; ep.?: Dodson 10783.

  [*Barbosella* sp. 1: Río Alambi, 2250 m, Molau & Ohman 1391.]

4. *Brachionidium* sp.
   Upper mont. for., Cerro Montecristi, 2750 m; terr., perianth purplish & green: 30558*.

5. *Brachtia andina* Rchb. f.
   Upper mont. for., 1800–2470 m; ep., fls. yellow: F 1380, 1479; P 3578; Croat 72873 (det. E. A. Christenson); Dalström 1604.

6. *Brassavola* sp.
   Prim. for., 1750–1800 m; ep., bracts red, fls. yellow: 27696.

7. *Brassia* cf. *arcuigera* Rchb. f.
   Sec. for., 1350 m; ep., fls. pale greenish mottled with brown: 31025.

8. *Brassia* sp. 1
   Sec. for., 1250 m; ep., fls. pale yellow: 31008.

9. *Chondrorhyncha embreei* Dodson & Neudecker
   Sec. for., 1200–2000 m; ep., fls. greenish-white with burgundy markings: 27648*, 28842*, 31173; C 5933*.

10. *Chondrorhyncha thienii* (Dodson) Dodson
    Prim. for., 2000 m; ep., fls. pale yellow, column purple-spotted: 28723*.

11. *Cleistes* sp.
    Sec. for., 1725 m; terr., fls. orange: 31925.

12. *Comparettia falcata* Poepp. & Endl.
    Sec. for., 1250 m; ep., fls. magenta: 31001.

    [*Cranichis antioquiensis* Schltr.: Nono–Nanegal, 2200 m, Sparre 16815.]

13. *Cranichis ciliata* (Kunth) Kunth
    Prim. for., Tandayapa, 2320 m; terr.: Luer 7305.

14. *Cranichis fertilis* (F. Lehm. & Kraenzl.) Schltr.
    Banks in sec. for., 1250–1800 m; terr., fls. white: 27166*, 27636*; V 12265; Harling & Andersson 11623; Todzia & Grimes 2479.

15. *Cranichis* sp. 1
    Sec. for., 1200–1300 m; terr., fls. whitish: 27783, 28840.

16. *Crossoglossa caulescens* (Lindl.) Dodson
    Nanegal, 1500 m: 1855, Jameson s.n.

17. *Crossoglossa nanegalensis* Dodson
    Sec. for., Tandayapa, 1700 m; terr., fls. yellow-green: Hirtz 2210*.

18. *Cryptocentrum* sp.
    Sec. for., 1550 m; ep. on mossy trunks: 31958.

19. *Cyclopogon inaequilaterus* (Poepp. & Endl.) Schltr.
    Sec. for., 1200–1350 m; terr., fls. yellowish-green: 27780*, 28109*.

20. *Cyclopogon ovalifolius* C. Presl
    Sec. for., 1300 m; terr., fls. white: 28783*.

21. *Cyclopogon* sp. 1
    Sec. for., 1300 m; terr., roots fleshy, fls. green with white lip: 27616.

22. *Cyclopogon* sp. 2
    Sec. for., 1300 m; terr., roots fleshy, fls. green with pink lip: 27617.

    [Two additional collections of *Cyclopogon*, 28725 and 29994, remain
    unidentified.]

23. *Dichaea longa* Schltr.
    Sec. for., Tandayapa, 1500 m; ep.: Dodson 16506.

24. *Dichaea morrisii* Fawc. & Rendle
    Sec. for., 1500–2200 m; ep.: 28020*, 28761*; Sparre 17045.

25. *Dichaea sodiroi* Schltr.
    Prim. & sec. for., 1500–2100 m; ep.: 27453*, 28874*, 28900*; C 5903*;
    1979, Luer s.n.

26. *Dichaea* sp. 1
    Sec. for., 1300–1400 m; ep., fls. yellow: 29301.

27. *Dichaea* sp. 2
    Prim. & sec. for., 1300–1775 m; ep.: 29273, 29579, 30391.

28. *Dracula dodsonii* (Luer) Luer
Nono–Tandayapa, km 17; ep.: Dodson 10792, 16501.

29. *Dracula navarroörum* Luer & Hirtz
Sec. for., Nanegalito, Tandayapa, 1600–1800 m; ep.: Dalström 1661; Hirtz 4854; Luer 15259.

30. *Dracula sodiroi* (Schltr.) Luer
Sec. for., 1750–2000 m; ep.: F 1199*; Dodson 15972.

31. *Dryadella* sp.
Prim. for., 2000 m; ep., fls. purplish: 31843.

32. *Elleanthus ampliflorus* Schltr.
Banks in sec. for., 1250 m; terr., fls. brick orange: 27301*.

33. *Elleanthus aurantiacus* (Lindl.) Rchb. f.
Sec. for., 2000 m; terr: K 1521.

34. *Elleanthus capitatus* (Poepp. & Endl.) Rchb. f.
Sec. for., 1300–2000 m; ep., stems to 1 m: 27534*, 27611*, 31547; C 5905*; L 11327; Bohlin 1100.

35. *Elleanthus discolor* (Rchb. f. & Warsz.) Rchb. f.
Sec. for., 1300 m; ep., bracts yellowish: 27708*.

36. *Elleanthus fractiflexus* Schltr.
Sec. for., 1400–1550 m; terr., fls. white: C 39783; Dodson 6982; Sodiro 35*.

37. *Elleanthus gastroglottis* Schltr.
Sec. for., 1800–2000 m; ep., perianth purple: C 5939*; Hurtado 1425*; Luer 2363.

38. *Elleanthus graminifolius* (Barb. Rodr.) Lojtnant
Sec. for., 1200–1250 m; ep.: 27531*, 28966*.

39. *Elleanthus linifolius* C. Presl
Sec. for., 1200–1250 m; ep.: 28967*.

[*Elleanthus petrogeiton* Schltr.: Calacalí–Nanegalito, K 1538.]

40. *Elleanthus robustus* (Rchb. f.) Rchb. f.
    Prim. for., 1650–1750 m; ep. or terr., stems to 1 m: 27447 ex p.*, 31565(?), 31917.

41. *Elleanthus smithii* Schltr.
    Banks, 1200–2000 m; terr., fls. orange, foetid: 28161*; Dodson & Thien 1107, Dodson 16505; Harling & Andersson 11613.

42. *Elleanthus vernicosus* Garay
    Sec. for., Tandayapa, 2000 m; terr.: Plowman & Davis 4446*.

43. *Elleanthus* sp. 1
    Sec. for., banks, 1550–1600 m; fls. white or magenta: 31119, 31122.

44. *Encyclia fragrans* (Sw.) Lemée [*Prosthechea fragrans* (Sw.) W. E. Higgins in CVPE]
    Sec. for., 1200–1250 m; ep., fls. white with red spots: 27535*.

45. *Encyclia hartwegii* (Lindl.) R. Vásquez & Dodson [*Prosthechea hartwegii* (Lindl.) W. E. Higgins in CVPE]
    Sec. for., 1300–2000 m; ep.: V 12309; Hirtz 1891*.

46. *Encyclia pamplonense* (Rchb. f.) Carnevali [*Prosthechea pamplonensis* (Rchb. f.) W. E. Higgins in CVPE]
    Tandayapa, 1800 m; ep.: Dodson 6989.

47. *Encyclia vespa* (Vell.) Dressler & G. E. Pollard [*Prosthechea vespa* (Vell.) W. E. Higgins in CVPE]
    Sec. for., 1700–2000 m; ep.: 28037*, 31643.

48. *Epidendrum arachnoglossum* Rchb. f. ex André
    Sec. scrub, above Río Pichán, 2050 m; terr., fls. white, lip with yellow spots: 30099*.

49. *Epidendrum arbuscula* Lindl.
    Prim. for., 1950–2000 m; ep., fls. green: 28715*.

50. *Epidendrum aristatum* Ackerman & Montalvo
    Prim. & sec. for., 1500–2250 m; ep. or terr.: 28004*, 28932*.

51. *Epidendrum brachyglossum* Lindl.
    Sec. for., 1800–2600 m: Dodson 6990; Molau & Eriksen 3057.

52. *Epidendrum brachystele* Schltr.
    Sec. for., Tandayapa, 2160 m: Dodson & Thien 1093, Dodson & Dodson 16502.

53. *Epidendrum brevilabre* Lindl.
    Upper mont. for., Cerro Montecristi, 2250 m; ep., fls. green with white lip: 28917*. [Not in CVPE.]

54. *Epidendrum calanthum* Rchb. f. & Warsz.
    Sec. for., 1250 m; ep., fls. yellow.: 28132*.

55. *Epidendrum caloglossum* Schltr.
    Sec. for., Nanegalito, 1600 m; ep.?: Dodson 6987.

56. *Epidendrum cochlidium* Lindl.
    Prim. & upper mont. scrub, 2000–2500 m; ep. or terr., fls. red or orange: 28085*, 28104*, 30260*, 30587*; F 1491; Asplund 17265; Dodson & Thien 1104.

57. *Epidendrum* cf. *cornanthera* Lehm. & Kraenzl.
    Sec. for., 1550 m; ep.?, viny: 31785.

58. *Epidendrum coryophorum* (Kunth) Rchb f.
    Loma Pahuamba, 2300 m; ep., fls. white with purple spots: F 1446*

59. *Epidendrum diothonaeoides* Schltr.
    Loma Pahuamba, 2300–2450 m; ep.: F 1350, 1434.

60. *Epidendrum elleanthoides* Schltr.
    Prim. for., 21 km west of Calacalí, 2500 m: L 13681.

61. *Epidendrum fimbriatum* Kunth
    Sec. for., 1800–2000 m; usually terr.: L 14046, 14048; Bohlin 1119; Harling & Andersson 11611; Dodson & Thien 1095.

    [*Epidendrum gastropodium* Rchb. f.: Calacalí–Nanegalito, K 1564; Nono–Tandayapa, km 16, 2200 m, Dodson 10764.]

62. *Epidendrum geminiflorum* Kunth
    Sec. & upper mont. for., 2000–2700 m; ep.: F 1413; K 1468; Dodson & Thien
    1100; Plowman & Davis 4436(?).

63. *Epidendrum goodspeedianum* A. D. Hawkes
    Prim. for., 1800–2100 m; ep., fls. green with pink lip: 28688*; Bohlin 1127;
    Dodson 10819.

64. *Epidendrum hymenodes* Lindl.
    Sec. for., Cerro Palo Seco, 1250–1300 m; ep., fls. green with white lip:
    27510*.

65. *Epidendrum* cf. *incomptum* Rchb. f.
    Sec. for., 1550–1600 m; ep., fls. greenish- or pinkish-brown: 31120, 32962.
    [Not in CVPE.]

66. *Epidendrum jamiesonis* Rchb. f. (incl. *E. evectum* Hook f.)
    Banks in sec. for., 2000 m; terr., to 1.5 m, fls. magenta: 31854; Dodson &
    Thien 1098; Gudiño 1409*.

67. *Epidendrum marsupiale* F. Lehm. & Kraenzl.
    Banks in sec. for., 1250 m; terr., fls. pale green: 31087**.

    [*Epidendrum pallatangae* Schltr.: Nono–Nanegal, 2200 m, Harling 14866.]

68. *Epidendrum paniculatum* Ruiz & Pav.
    Sec. for., Nono–Nanegal, 1600–2000 m; terr.: Dodson & Thien 1094, Dodson
    6986.

69. *Epidendrum peltatum* Schltr.
    Prim. for., 2000 m.; ep., fls. magenta: 28861*.

70. *Epidendrum* cf. *porphyreum* Lindl.
    Banks, 1600–2450 m; ep. or terr., fls. pink to magenta: 31918; F 1334.

71. *Epidendrum ramosum* Jacq.
    Sec. for., 1200–1500 m; ep.: 28968*, 32916; Dodson 16509.

72. *Epidendrum renilabium* Schltr.
    Sec. for., 1700–2000 m; ep., fls. yellowish: 27951*, 31924.

[*Epidendrum repens* Cogn.: Nono–Tandayapa, km 16, Dodson 10768.]

73. *Epidendrum* cf. *secundum* Jacq.
Sec. for. & banks, 1200–1875 m; ep. or terr., fls. orange: 27502*, 28051*, 28153, 31631.

74. *Epidendrum spathatum* Schltr. (incl. *E. embreei* Dodson)
Sec. for. & scrub, 1650–2300 m; ep. or terr., fls. orange: 30067*, 30098*, 31181**, 31792; F 1068, 1402; P 3589*; Croat 72862*; Dodson & Thien 1090; Hurtado 1437*, 1441*.

75. *Epidendrum tandapianum* Dodson & Hágsater
Prim. for., 2200 m; ep., fls. green: 28931*.

76. *Epidendrum* cf. *tunguraguae* Schltr.
Banks, 1200–1650 m; terr., stems to 2 m, fls. greenish-white with creamy white lip: 28163*, 28389, 31922; Holm-Nielsen 24507*.

77. *Epidendrum* sp. 1
Sec. for., 1200 m; ep.: 28843*.

78. *Epidendrum* sp. 2
Sec. for., 1300 m; ep., fls. pale green: 31111.

79. *Erythrodes boliviensis* Cogn.
Prim. & sec. for., 1650–2000 m; terr., fls. pinkish: 28223*, 28858*, 30066*. [Not in CVPE.]

80. *Erythrodes erythrodoides* (Schltr.) Ames
Sec. for., 1450 m; terr., fls. greenish: 28849.

81. *Erythrodes jamesonii* (Garay) Dodson
Nanegal, 1300 m; terr.: Jameson s.n.

[A number of additional collections of *Erythrodes* remain unidentified to species.]

82. *Eurystyles cotyledon* Wawra
Sec. for., 1450–1500 m; ep.: 31905.

83. *Gomphichis hetaerioides* Schltr.
    Sec. for., Nanegal to Gualea, 1300 m; terr.: Sodiro s.n. [apparently right at the boundary of our area].

84. *Govenia sodiroi* Schltr.
    Between Nono & Tandayapa, 2100 m: Harling 19976.

85. *Habenaria dentifera* C. Schweinf.
    Sec. for., 1350–1400 m; terr., fls. greenish: 27593*, 32359.

86. *Habenaria floribunda* Lindl.
    Sec. for., Nanegal; terr.: Sodiro s.n.*.

87. *Habenaria monorrhiza* (Sw.) Rchb. f.
    Sec. for., 1200–1850 m; terr., fls. white or greenish: 27089*, 27781*, 27785*, 28038, 28055*, 28164*, 28264*; Dodson 10824; Harling & Andersson 11560; Todzia 2478.

88. *Isochilus linearis* (Jacq.) R. Br.
    Sec. for., 1300–1350 m; ep.: 32360.

89. *Kefersteinia sanguinolenta* Rchb. f.
    Sec. for., Tandayapa, 1600 m; ep.: Dodson 18799A.

90. *Kefersteinia* sp. 1
    Rip. for., 1900 m; ep.: 31592.

    [*Lepanthes brachypodon* Luer: west of Tandayapa, 2320 m, Luer 7301 (type collection).]

91. *Lepanthes casidea* Rchb. f.
    Forests, Nanegal; ep.: 1854, Jameson.

92. *Lepanthes effusa* Schltr.
    Nono–Nanegal, 2000 m: Dodson & Thien 1106.

93. *Lepanthes gargantua* Rchb. f.
    Prim. for., 1550–1600 m; ep., lvs. purplish, fls. white: 31055.

94. *Lepanthes mucronata* Lindl.
    Forests, Nanegal; ep.: 1854, Jameson.

95. *Lepanthes nanegalensis* Rchb. f.
Nanegal, ep.: "1864" [1854?], Jameson.

96. *Lepanthes pecunialis* Luer
Sec. for., 1250–1600 m; ep., fls. orange: 27613*, 28130*; T 595; Dalström 1609; Luer 4753.

97. *Lepanthes pilosella* Rchb. f.
Nono–Tandayapa, km 17: Dodson 10788.

98. *Lepanthes rhodophylla* Schltr.
Nanegal Valley near Río Frio; ep.: 1902, Sodiro 1896.

[Luer, 1986, reports 5 additional species of *Lepanthes* from near La Liberia & Pellagallo, just east of our boundary: *L. acarina* Luer; *L. biloba* Lindl.; *L. columbae* Luer; *L. pteropogon* Rchb. f.; and *L. stupenda* Luer.]

99. *Liparis "harlingii"*
Sec. for., Nanegal, 1200 m; terr.: Harling & Andersson 11599*. [Not in CVPE; unpublished name.]

100. *Lockhartia longifolia* (Lindl.) Schltr.
Sec. for., 1250–1600 m; ep., fls. yellow: 27625*, 28964*, 30428*.

101. *Lycaste grande* Oakeley & Fowlie
Sec. for., Nanegal, 1800 m: Dalström 1606.

102. *Lycaste* cf. *longipetala* (Ruiz & Pav.) Garay
Rip. for., 2000 m; ep., fls. yellowish: 31819.

103. *Lycomormium* sp.
Sec. for., Inca Trail, 1350 m; ep., buds whitish: 31083.

104. *Macroclinium perryi* (Dodson) Dodson
Sec. for., Nanegalito, 1400–1600 m; ep.?: Dodson 6995 [type collection]; Hirtz 993.

105. *Masdevallia anachaeta* Rchb. f.
Nono–Tandayapa, km 17: Dodson 10787.

106. *Masdevallia angulata* Rchb. f.
Sec. for., 1500–2000 m; ep.: C 5938; Luer 4723*.

107. *Masdevallia nidifica* Rchb. f.
Sec. for., 1900–2000 m; ep., fl. cream & purple: F 1020; Luer 2371*.

108. *Masdevallia ophioglossa* Rchb. f.
Sec. for., 1700 m; ep.: Luer 5211*.

109. *Masdevallia parvula* Schltr.
Nono–Nanegal, 2000 m; ep.: Dodson 1096.

110. *Masdevallia* cf. *ventricularia* Rchb. f.
Prim. for., 1675–2250 m; ep., fls. maroon with orange tips: 28948, 31136; Luer 1695.

111. *Masdevallia* sp. 1
Sec. for., 2100–2200 m; ep., fls. yellow: 30088.

112. *Masdevallia* sp. 2
Prim. for., 2000 m; ep., fls. greenish-white: 28860.

113. *Maxillaria acutifolia* Lindl.
Sec. for., 1200–1250 m; ep., fls. orange with burgundy lip: 27512*, 27536*.

114. *Maxillaria aurea* (Poepp. & Endl.) L. O. Williams
Loma Pahuamba, 1900–1950 m; ep.: F 1146.

115. *Maxillaria cryptobulbon* Carnevali & J. T. Atwood
Sec. for., 1375–1500 m; ep., fls. white or yellowish: 30396*, 31872, 32924.

116. *Maxillaria ecuadorensis* Schltr.
Sec. for., 1200–2100 m; ep. or terr., fls. yellow with reddish-brown lobes: F 1249, 1495; N 8655*; Hurtado 1420*.

117. *Maxillaria grandiflora* (Kunth) Lindl. (incl. *M. lehmannii* Rchb. f.)
Sec. & upper mont. for., 1600–2400 m; ep., perianth white, lip yellow within, purple-striped without: 32815; Dodson 15749.

118. *Maxillaria jamesonii* (Rchb. f.) Garay & Schweinf.
Nono–Tandayapa, 2200 m: Harling & Andersson 11633.

119. *Maxillaria lepidota* Lindl.
   Prim. & sec. for., 1600–2200 m; ep., fls. yellow with purple spots: 30457*, 31080, 32839; P 3579*; Dalström 1605; Dodson & Thien 1089.

120. *Maxillaria nanegalensis* Rchb. f.
   Nanegal; ep.: Jameson s.n.

121. *Maxillaria nigrescens* Lindl.
   Nono–Nanegal, 2000 m: Dodson & Thien 1099.

122. *Maxillaria pardalina* Garay
   Nono–Nanegal and Loma Pahuamba, 2000–2100 m; ep., fls. yellow with red spots: F 1493; Dodson 1091.

123. *Maxillaria parviflora* (Poepp. & Endl.) Garay
   Sec. for., 1100–1225 m; ep.: 31160; Dodson 17063.

124. *Maxillaria. polyphylla* Rchb. f.
   Nono–Nanegal, 2000 m: Dodson & Thien 1092.

125. *Maxillaria porrecta* Lindl.
   Sec. for., 1200–1300 m; ep., fls. yellowish: 27511*, 27532*, 27610*.

126. *Maxillaria ramosa* Ruiz & Pav.
   Sec. for., 1250–1300 m; ep., viney, fls. green: 27615*.

[Two species of *Maxillaria* collected from Nono to Nanegal at 2100–2200 m appear to be slightly outside our limits: *M. luteo-rubra* (Lindl.) Rchb. f., Luer 4710; Plowman & Davis 4443; and *M. squarrosa* (Schltr.) Dodson, Dodson 16504.]

127. *Myoxanthus* sp. 1
   Sec. for., 1600–1900 m; ep., fls. pale yellow: 27821, 27966, 28010.

128. *Myoxanthus* sp. 2
   Sec. for. & banks, 1250–1350 m; ep. or terr., fls. greenish: 31031, 31088.

129. *Myrosmodes* sp.
   Prim. for., 1900 m; ep.: 30458*.

130. *Odontoglossum cirrhosum* Lindl.
   Scrub on banks, above Río Pichán, 2000 m; terr., fls. white with red dots, lip yellow: 30589**; Luer 2373.

131. *Odontoglossum cristatum* Lindl.
   Sec. for., 1850 m; ep., fls. yellow & brown: 30382**; Dodson 16507.

132. *Odontoglossum "denticulatum"*
   Prim. & upper mont. for., 2000–2750 m; ep., fls. yellow & brown: 30556**; Andreetta 216; Lehmann 8549. [Not in CVPE; unpublished name.]

133. *Odontoglossum hallii* Lindl.
   Prim. for., 1800–2000 m; ep., fls. yellow-green spotted with brown: 27333*.

   [*Oncidium "geniculatum"*, Nono–Nanegal, 2200 m: Hirtz 1041 (unpublished name).]

134. *Oncidium hapalotyle* Schltr.
   Sec. for., Tandayapa, 1900 m: Harling & Andersson 11615*.

135. *Oncidium hartwegii* Lindl.
   Prim. for., 2050–2275 m; ep., fls. brown & yellow: 28934*, 29213, 29450, 29551.

136. *Oncidium heteranthum* Poepp. & Endl.
   Nono–Tandayapa, km 17; ep.: Dodson 10784.

137. *Oncidium klotzscheanum* Rchb. f. (incl. *O. obryzatum* Rchb. f.)
   Sec. for., 1300 m; ep., fls. yellow with brown spots on lateral tepals: 27619*, 32929.

138. *Oncidium meirax* Rchb. f.
   Prim. for., 1200–2100 m; ep., fls. dull yellow with brown spots: 27447 ex p.*, 27450*, 28132*, 28951*; T 177*; Filskov 37131; Holm-Nielsen 24504.

139. *Oncidium orthotis* Rchb. f.
   Sec. for., 1300–1600 m; ep., fls. yellow with brown spots: 27937*; Dodson 16510.

140. *Oncidium pentadactylon* Lindl.
Prim. & upper mont. for. or scrub, 1550–2500 m; ep. or terr., fls. yellow: 28093*; K 1487; Dodson 6991; Holm-Nielsen 24505. "Margarita"

141. *Oncidium serratum* Lindl.
Prim. & sec. for. and clearings, 1500–2225 m; common, ep. or terr., inflors. to 4 m, twining or clambering, fls. yellow & brown: 27960*, 28058*, 28891*, 29402, 30100*, 30590; Dodson & Thien 1102; Holmgren 851; Holm-Nielsen 24484. "Margarita"

142. *Pachyphyllum falcifolium* Rchb. f.
Calacalí–Nanegal: K 1523.

143. *Pelexia ecuadorensis* Schltr.
Nanegal, 1300 m; terr.: Jameson s.n.

144. *Peristeria* cf. *lindenii* Rolfe
Sec. for., 1350 m; ep., buds white: 31024.

145. *Phragmipedium lindenii* (Lindl.) Dressler & N. Williams
Sec. for., 1300–1700 m; terr., tepals yellow with greenish veins, lobes purplish: 27968*; Dodson 10821; Sodiro s.n.

146. *Phragmipedium longifolium* (Rchb. f. & Warsz.) Rolfe
Sec. for., Nanegal; terr.: Hartweg s.n.*.

[*Platystele alucitae* Luer, Dodson 7315; and *Platystele microscopa* Luer, Dodson 7311; both from Tandayapa–Mindo, 2100 m.]

147. *Platystele* sp.
Prim. for., 1650–1700 m; ep., fls. greenish: 31143.

148. *Pleurothallis anceps* Luer
Sec. for., Nanegalito, 1600 m; ep.: Dodson 6993.

[*Pleurothallis cassidis* Lindl.: at Hacienda Yunguilla, 2800 m, Haught 3181.]

149. *Pleurothallis chloroleuca* Lindl.
Sec. for., 1500–1700 m; ep., fls. yellow: Q 121 (det. G. Carnevali).

150. *Pleurothallis cordata* (Ruiz & Pav.) Lindl.
    Prim. & upper mont. for., 1700–2725 m; ep., plants greenish or purplish: 27443, 28206*, 28253*, 29502, 30560, 30567, 31227; L 13678, 13685, 13688.

151. *Pleurothallis imperialis* Luer
    Banks between El Carmen & Marianitas, 1200–1250 m; ep.: 28159*.

152. *Pleurothallis* cf. *macra* Lindl.
    Upper mont. for., 2500–2750 m; ep., fls. yellow: K 31979; L 13675, 13682, 13683.

153. *Pleurothallis ramulosa* Lindl.
    Nono–Nanegal, 2000 m; ep.: Dodson & Thien 1103.

154. *Pleurothallis ruscifolia* (Jacq.) R. Br.
    Prim. for., 1700–1800 m; ep., common, fls. yellow: 27811*, 28061*, 28062, 28069*, 28194*.

155. *Pleurothallis scabrilinguis* Lindl.
    Sec. for., 1800 m; ep.: Luer 13321.

156. *Pleurothallis sclerophylla* Lindl.
    Loma Pahuamba, 1900–1950 m; ep.: F 1141.

157. *Pleurothallis sicaria* Lindl.
    Sec. for., 1350–1400 m; ep., fls. & frs. dark green: 32918 (det. D. Kelch).

158. *Pleurothallis stevensonii* Luer
    Prim. for., 1750 m; ep., fls. white with red spots: 28846*.

159. *Pleurothallis undulata* Poepp. & Endl.
    Sec. for., Tandayapa, 1200–1300 m; ep.: Hirtz 1892, 1893.

160. *Pleurothallis zephyrina* Rchb. f.
    Sec. for., 2150 m; ep.: Dodson 15750.

[A considerable number of collections of *Pleurothallis* remain unidentified to species.]

161. *Polystachya* cf. *foliosa* (Lindl.) Rchb. f.
    Sec. for., 1250–1600 m; ep., fls. translucent yellow: 30423, 31567.

162. *Ponthieva disema* Schltr.
    Sec. & upper mont. for., 2000–2600 m; terr., fls. white: 32402; Haught 3162.

163. *Ponthieva pseudoracemosa* Garay
    Sec. for., 1600–2200 m; terr.: Smith 1946; Sparre 16795.

164. *Porphyrostachys* sp.
    Sec. for., 1300–1400 m; terr., fls. green: 32358. [Our plants differ from *Porphyrostachys pilifera* (Kunth) Reichenb. f. in the distinctly green, rather than red, flowers.]

165. *Porroglossum amethystinum* (Rchb. f.) Garay
    Sec. for., Cerro Negro, 1500–2000 m; ep., fls. purplish: 27963*.

166. *Porroglossum muscosum* (Rchb. f.) Schltr.
    Sec. for., 2000 m; ep.: C 5952.

167. *Porroglossum* sp. 1
    Sec. for., 1475 m; ep. on mossy log, fls. pale maroon: 31898.

168. *Prescottia stachyodes* (Sw.) Lindl.
    Sec. for., 1300–2000 m; terr., fls. greenish-white: 32384, 32455; Hirtz 1890.

169. *Pseudocentrum macrostachyum* Lindl.
    Sec. for., 2200–2650 m; terr., fls. green: Dodson 15774; Molau 3037.

170. *Psygmorchis pumilio* (Rchb. f.) Dodson & Dressler
    Sec. for., 1250 m; ep., fls. yellow: 28129*.

171. *Restrepiopsis tubulosa* (Lindl.) Luer
    Nono–Tandayapa, km 17; ep.: Dodson 10786.

172. *Rodriguezia lehmannii* Rchb. f.
    Sec. for., 1250–1700 m; ep., fls. white with purplish or reddish lines: 27447 ex p.*, 27845*, 27977*, 31109; T 597*. [29078, with yellowish-green fls., may represent another species.]

173. *Scaphosepalum* sp.
    Prim. & upper mont. for., 2250–2700 m; ep., fls. purple: 28940*; K 31983.

174. *Scelochilus chiribogae* Dodson
Sec. for., Nanegalito, 1400 m; ep.: Hirtz 989*.

175. *Scelochilus heterophyllus* Rchb. f.
Sec. for., Tandayapa, 1800 m; ep.: Hirtz 2179*.

176. *Scelochilus jamesonii* Lindl.
Sec. for., Tandayapa, 2000 m; ep.: Dodson 16653.

177. *Sertifera purpurea* Lindl. & Rchb. f.
Prim. for., 21 km west of Calacalí, 2500 m: L 13677.

178. *Sertifera* sp. 1
Prim. for., 2250–2300 m; ep.: 28919*.

[*Sigmatostalix adamsii* Dodson: Quito–Tandayapa, 2300 m, Dodson 13093.]

179. *Sigmatostalix* sp.
Sec. for., 1300 m; ep., fls. dull yellow with brown spots: 27684.

180. *Sobralia crocea* (Poepp. & Endl.) Rchb. f.
Sec. for., 1600 m; terr.: Dodson 6992; Molau 2230.

181. *Sobralia "ecuadorana"* Dodson (ined.)
Sec. for., Tandayapa, 1850 m; terr.: Dodson 10820.

182. *Sobralia gloriosa* Rchb. f.
Sec. for., 1200–1500 m; terr., stems 1 m, fls. white with maroon lip: V 12264*.

183. *Sobralia klotzscheana* Rchb. f.
Sec. for., 1450–2100 m; terr., to 1.5 m, fls. pink or white with pink lip: 27967*, 30451, 30481(?), 30502, 32969. "Maygua"

184. *Sobralia pulcherrima* Garay
Banks, 1200–1800 m; terr., to 2 m, fls. pink, lip with purplish lines: 28302, 29979*, 31941; Harling & Andersson 11751 (between Nanegal & Nanegalito, [type collection]); Holm-Nielsen 24516*.

185. *Sobralia rosea* Poepp. & Endl.
Steep banks in sec. for., 1200–1300 m; terr., canes to 3 m, fls. white, lip with purplish lines: 27079*; Asplund 1571; Sodiro 141.

186. *Sobralia valida* Rolfe
Sec. for., 1300 m; terr., canes to 1 m; fls. fragrant, white with yellow throat, lip violet-edged: 31342**.

187. *Stanhopea impressa* Rolfe
Sec. for., Tandayapa, 1600 m; ep.: Dodson 18799.

188. *Stanhopea* sp. 1
Sec. for., Hacienda El Carmen, 1250 m; ep.: Osbourn s.n.

[*Stelis alba* Kunth: at Hacienda Yunguilla, 2800 m, Haught 3187.]

189. *Stelis morganii* Dodson & Garay
Sec. for., 1550–1600 m; ep., fls. cream: 31784; T 596*.

190. *Stelis pusilla* Kunth
Prim. for., 1650–2000 m; ep., fls. pale yellow: 27964*; L 14045.

191. *Stelis vulcanica* Rchb. f.
Nono–Nanegal, 2000 m: Luer 6325A.

[A number of additional collections of *Stelis* remain unidentified to species.]

192. *Stellilabium andinum* (L. O. Williams) Garay & Dunst.
Sec. for., Nanegalito, 1400 m; ep.: Hirtz 991*.

193. *Stellilabium astroglossum* (Rchb. f.) Schltr. (incl. *S. tanii* Dodson)
Sec. for., Tandayapa, Nanegalito 1500–1600 m; ep.: Dodson 6994, 6995, 16511.

194. *Stellilabium hirtzii* Dodson
Sec. for., Nanegalito, 1400 m; ep.: Hirtz 992* [type collection].

[*Stenorrhynchos cernuus* Lindl., west of Nanegal, 1200 m, Hartweg s.n. (type collection) is probably extralimital.]

195. *Symphyglossum sanguineum* (Rchb. f.) Schltr.
Sec. for. & scrub, 1700–2000 m; ep., fls. red: 30155 (det. S. Dalstrom), 30588*; C 7175*; Dodson 6984.

196. *Trichopilia fragrans* Lindl.
   Prim. for., 1300–2000 m; ep. or terr., fls. white, fragrant: 32920; C 5926*.

197. *Trichopilia rostrata* Rchb. f.
   Prim. for., 1750 m; ep., fls. white, throat pink within: 28848*.

198. *Trichosalpinx* dura (Lindl.) Luer
   Prim. for., 1900–2000 m; ep., buds yellowish: 27751; Dodson & Thien 1105.

199. *Trichosalpinx* sp. 1
   Sec. for., 1600 m; ep.: 31124.

200. *Trisetella vittata* (Luer) Luer
   Sec. for., Tandayapa; ep.: Luer 5209*.

201. *Xylobium elongatum* (Lindl. & Paxt.) Hemsl.
   Prim. & sec. for., 1250–1625 m; ep., fls. whitish: 28390, 31074.

202. *Xylobium leontoglossum* (Rchb. f.) Rolfe
   Sec. for., 1200–2150 m; ep., fls. dull creamy white: 30368*, 31009, 31152(?);
   C 5927*; T 156*; Hurtado 1436*; Molau 3041.

203. *Xylobium pallidiflorum* (Hook.) Nicholson
   Sec. for., 1250–1600 m; ep., fls. pink: 27933*, 28175*; T 594*.

204. *Zootrophion hirtzii* Luer
   Sec. for., 1600 m; ep.: Hirtz 2166.

POACEAE (31) (* det. G. Davidse; ** det. S. Laegaard; *** det. S. Renvoize)
   Refs.: Hitchcock, A. S. 1927. Contr. U. S. Nat. Herb. 24: 291–556; Judziewicz,
   E. J. 1990. Fl. Guianas A8: 1–727.

1. *Aegopogon cenchroides* Humb. & Bonpl. ex Willd.
   Open scrub, 2000 m: 32845.

2. *Agrostis* sp.
   Rocky slope in sec. for., 1850–1900 m; herb with sprawling stems: 31575.

3. *Andropogon bicornis* L.
   Sec. for. & cleared areas, 1200–1650 m; common: 27159, 28354, 29353; Z 63.
   "Puntero"

4. *Andropogon leucostachyus* Kunth
   Disturbed roadsides, 1850–1900 m: 31635.

5. Anthoxanthum odoratum L.
   Bed of Río Pichán, 1900 m: 32396.

6. *Arundinella berteroniana* (Schult.) A. Hitchc. & Chase
   Cliffs above Río Umachaca, 1400 m; culms 1.5 m: 27156**. [Record doubtful; a duplicate of this number was determined by Davidse as *Setaria sphacelata*.]

7. *Axonopus compressus* (Sw.) P. Beauv. [upland form]
   Sec. for., 1300–1400 m: 27157*, 27869*.

8. *Axonopus scoparius* (Flüggé) Kuhlm.
   Pastures & sec. scrub, 1300–1400 m; robust grass 1–1.5 m high: 31703; Filskov 37050; Z 60, 199. "Gamalote" [An important forage grass.]

9. Briza minor L.
   Mossy banks, 2000 m: 31859.

10. *Chusquea scandens* Kunth
    Sec. for., 1800–2200 m; climber: C 13055; F 1471.

11. *Chusquea uniflora* Steud.
    Loma Pahuamba, 2300 m; climber: F 1422.

    [Additional collections of *Chusquea* from primary forest, Cerro Sosa, 2050–2650 m, climbing to 8 m: 29138, 29179, 29434, 29487, all barren, remain to be identified.]

12. Coix lacryma-jobi L.
    Sec. for. & scrub, 1250 m: 28303.

13. *Cortaderia bifida* Pilger
    Scrub on banks, west of Quebrada Chiquilpe, 1950–2200 m; leaves 1–1.5 m, inflors. 2–2.5 m: 32844.

14. Cymbopogon citratus (DC.) Stapf
    Sec. for., 1200–1400 m: Filskov 37072***.

15. <u>Digitaria</u> <u>abyssinica</u> (Hochst. ex A. Rich.) Stapf
    Sec. for., 1500 m: Z 66*.

16. <u>Eleusine</u> <u>indica</u> (L.) Gaertn.
    Pastures, 1300–1500 m: Z 156, 223.

17. *Guadua angustifolia* Kunth
    Common along Río Alambi, 1200–1300 m; canes to 10 m: 32863.

18. *Gynerium sagittatum* (Aubl.) P. Beauv.
    Banks of Río Alambi, 1125 m; culms 3–4 m high, spikes pendent, purplish:
    31279.

19. <u>Holcus</u> <u>lanatus</u> L.
    Openings in upper mont. for., 2700–2790 m: 30542.

20. *Homolepis glutinosa* (Sw.) Zuloaga & Soderstrom
    Sec. for., 1175 m: 28809*.

21. *Ichnanthus* sp.
    Clearings (potreros) in sec. for., 1250–1300 m: 31878.   [Resembles
    I. candicans (Nees) Döll, which is not listed in CVPE.]

22. *Isachne arundinacea* (Sw.) Griseb.
    Sec. for., 1225 m; vine with pendent branches: 31093.

23. *Lasiacis nigra* Davidse
    Sec. for., 1300–1700 m; common scrambler: 27394, 27440, 27563, 27971,
    29304*, 31039, 31657; M 5254*.

24. *Muhlenbergia* cf. *tenuifolia* (Kunth) Kunth
    Rocky slopes in sec. for., 1900 m: 31625. [Not in CVPE.]

25. <u>Oplismenus</u> <u>burmannii</u> (Retz.) P. Beauv.
    Clearings in sec. for., 1900–1950 m: 31620.

26. *Oplismenus hirtellus* (L.) P. Beauv.
    Weed in fields, sec. for., clearings in prim. for., 1150–1700 m: 27190, 27842,
    28835; Z 86.

27. *Panicum hebotes* Trin.
   Sec. for., 1250–1400 m; weed: 27838*, 27913*.

28. *Panicum laxum* Sw.
   Open areas, 1400 m: 27154.

29. *Panicum maximum* Jacq.
   Sec. for., 1400 m: Filskov 37051.

30. *Panicum polygonatum* Schrad.
   Weed in banana fields, 1300–1500 m: 27200; Filskov 37081**.

31. *Panicum pulchellum* Raddi
   Sec. for., 1350–1400 m: 30288.

32. *Panicum* cf. *sellowii* Nees
   Sec. for., 1450 m: 29041.

33. *Paspalum conjugatum* Bergius
   Roadsides & fields, 1200–1350 m: 27158; Z 170.

34. *Paspalum* cf. *decumbens* Sw.
   Pastures, 1300–1500 m: Z 159.

35. *Paspalum paniculatum* L.
   Weed in fields, 1400–1500 m: 27203; Z 59.

36. *Paspalum penicillatum* Hook. f.
   Prim. for., 2400 m: 32802.

37. *Paspalum saccharoides* Nees ex Trin.
   Roadside banks, 1200–1250 m: 27163, 31102.

38. *Pennisetum polystachion* (L.) Schult.
   Roadside banks, 1200–1250 m: 27293. [Listed as *Pennisetum setosum* (Sw.) Rich. in CVPE.]

39. Pennisetum purpureum Schumach.
   Grassy slopes along Río Alambi, 1135 m; stems 2 m: 31274. "Pasto elefante"

40. *Pennisetum tristachyum* (Kunth) Spreng.
   Streambeds & roadsides, above Río Pichán, 2000–2225 m: 30101, 32424.
   "Tundillo"

41. Poa annua L.
   Rip. vegetation, 1600–1900 m: 32395, 32789.

42. *Pseudechinolaena polystachya* (Kunth) Stapf
   Sec. for. & pastures, 1200–1400 m; common: 27013, 27768, 27885, 28114**,
   29070; Z 16; Filskov 37120***.

43. *Setaria parviflora* (Poir.) Kerguélen
   Mossy banks in prim. for., 1400–2000 m: 31858; Filskov 37082***.

44. Setaria sphacelata (Schumach.) M. B. Moss ex Stapf & C. E. Hubb.
   Sec. rip. for., 1200–1400 m: 27156*, 28112*; Z 61, 188; Filskov 37046.
   "Miel"

45. *Sporobolus indicus* (L.) R. Br.
   Open areas, 1200–1300 m: 27160, 29553; Z 64.

46. Vulpia myuros (L.) C. C. Gmel.
   Mossy banks, 2000 m: 31860.

47. *Zeugites mexicana* (Kunth) Trin. ex Steud.
   Mature sec. for., 1400–1900 m; common herb with clumped stems, petiolate
   leaves: 31052, 32374, 32392; Laegaard 71453.

PONTEDERIACEAE (1)
   Ref.: Horn, C. N. 1987. Fl. Ecuador 29: 1–19.

1. *Heteranthera reniformis* Ruiz & Pav.
   Muddy flats, 1250–1700 m: 27383.

SMILACACEAE (1)
   Ref.: Killip, E. P., & C. V. Morton. 1936. Publ. Carnegie Inst. Wash. 461:
   257–290.

1. *Smilax* cf. *eucalyptifolia* Kunth
   Sec. for., 1500–1650 m; woody vine with yellowish-green fls.: 27122, 29351,
   31128, 31910. "Zarzaparilla" [Not in CVPE.]

2. *Smilax kunthii* Killip & C. V. Morton
   Prim. for., 1800–2100 m; vine: 28933, 28953; C 5908, 7189 (last two det. J. Gaskin).

TYPHACEAE (1)
Ref.: Crespo, S., & R. L. Pérez-Moreau. 1967. Darwiniana 14: 413–429.

1. *Typha domingensis* Pers.
   Along Río Umachaca, 1200–1250 m: 28974.

ZINGIBERACEAE (2) (* det. C. Ulloa; ** det. P. J. M. Maas)
Ref.: Maas, P. J. M. 1976. Fl. Ecuador 6: 1–49.

1. Hedychium coronarium J. König
   Roadsides, 1200–1300 m; fls. white: 27802.

2. *Renealmia aurantifera* Maas
   Sec. for., Cerro Negro, 1500–2175 m; bracts greenish, fls. orange: 28009*, 30116, 30487; Croat 72868**.

3. *Renealmia dolichocalyx* Maas
   Sec. for., 1250–2050 m; stems 2–2.5 m, inflors. separate from base, fls. white, frs. red: 28747, 30506, 31251, 31339, 31389, 31737.

4. *Renealmia fragilis* Maas
   Loma Pahuamba, 2200 m; stems to 0.8 m, fls white, imm. frs. red: F 1475**.

5. *Renealmia sessilifolia* Gagnep.
   Sec. for., 1250–1700 m; to 2 m, bracts red, fls. white: 27101, 27149, 27672, 27865, 29350, 30034, 31069, 31662; G 73197**; V 12320**.

6. *Renealmia thyrsoidea* (Ruiz & Pav.) Poepp. & Endl.
   Sec. for., 1200–1250 m; stems to 2 m, inflors. separate from base, bracts red: 28544*. "San Juanillo"

7. *Renealmia* sp. 1
   Prim. for., 1975–2275 m; stems c. 1 m, inflors. separate from base, fls. white with orange center, frs. dark red: 29150, 29227. [Possibly a form of *R. oligosperma* K. Schum.]

Appendix A. Dominant or conspicuous taxa of secondary and roadside vegetation in the Maquipucuna region, between 1150 and 1500 m

Trees and arborescent shrubs:
Cyathea caracasana (Cyatheaceae)
Saurauia cf. *prainiana* (Actinidiaceae)
Baccharis spp. (Asteraceae)
Begonia parviflora (Begoniaceae)
Delostoma integrifolium (Bignoniaceae)
Ceiba aff. *salmonea* (Bombacaceae)
Cordia aff. *spinescens* (Boraginaceae)
Tournefortia spp. (Boraginaceae)
Cecropia monostachya (Cecropiaceae)
Acalypha diversifolia (Euphorbiaceae)
Banara guianensis (Flacourtiaceae)
Besleria angustiflora (Gesneriaceae)
Wercklea ferox (Malvaceae)
Siparuna aspera (Monimiaceae)
Otoba gordoniifolia (Myristicaceae)
Psidium guajava (Myrtaceae)
Palicourea spp. (Rubiaceae)
Psychotria spp. (Rubiaceae)
Acnistus arborescens (Solanaceae)
Triumfetta grandiflora (Tiliaceae)
Trema micrantha (Ulmaceae)

Lianas and vines:
Iresine diffusa (Amaranthaceae)
Mandevilla callista (Apocynaceae)
Matelea spp. (Asclepiadaceae)
Mikania micrantha (Asteraceae)
Ipomoea alba (Convolvulaceae)
Lagenaria siceraria (Cucurbitaceae)
Melothria pendula (Cucurbitaceae)
Desmodium campyloclados (Fabaceae)
Phaseolus coccineus (Fabaceae)
Hydrangea peruviana (Hydrangeaceae)

Cissampelos tropaeolifolia (Menispermaceae)
Rubus sp. (Rosaceae)
Solanum juglandifolium (Solanaceae)
Cissus verticillata (Vitaceae)
Vitis tiliifolia (Vitaceae)

Weedy understory herbs, subshrubs, and pteridophytes:
Sticherus bifidus (Gleicheniaceae)
Thelypteris spp. (Thelypteridaceae)
Cyathula achyranthoides (Amaranthaceae)
Eryngium foetidum (Apiaceae)
Sanicula liberta (Apiaceae)
Asclepias curassavica (Asclepiadaceae)
Bidens pilosa (Asteraceae)
Erato polymnioides (Asteraceae)
Munnozia spp. (Asteraceae)
Polyanthina nemorosa (Asteraceae)
Begonia foliosa (Begoniaceae)
Heliotropium rufipilum (Boraginaceae)
Tournefortia spp. (Boraginaceae)
Drymaria cordata (Caryophyllaceae)
Crotalaria sagittalis (Fabaceae)
Desmodium spp. (Fabaceae)
Phaseolus spp. (Fabaceae)
Irlbachia alata (Gentianaceae)
Columnea picta (Gesneriaceae)
Hyptis spp. (Lamiaceae)
Nasa triphylla (Loasaceae)
Cuphea spp. (Lythraceae)
Pavonia castaneifolia (Malvaceae)
Sida acuta (Malvaceae)
Urena lobata (Malvaceae)
Tibouchina longifolia (Melastomataceae)

197

Fuchsia macrostigma (Onagraceae)
Ludwigia octovalvis (Onagraceae)
Bocconia integrifolia (Papaveraceae)
Phytolacca rivinoides (Phytolaccaceae)
Piper spp. (Piperaceae)
Monnina spp. (Polygalaceae)
Polygala paniculata (Polygalaceae)
Borreria spp. (Rubiaceae)
Scoparia dulcis (Scrophulariaceae)
Browallia americana (Solanaceae)
Capsicum lycianthoides (Solanaceae)
Jaltomata procumbens (Solanaceae)
Boehmeria spp. (Urticaceae)
Urera baccifera (Urticaceae)
Valeriana chærophylloides (Valerianaceae)
Lantana camara (Verbenaceae)
Stachytarpheta cayennensis (Verbenaceae)
Verbena litoralis (Verbenaceae)
Tradescantia zanonia (Commelinaceae)
Tripogandra serrulata (Commelinaceae)
Costus spp. (Costaceae)
Cyperus spp. (Cyperaceae)
Heliconia spp. (Heliconiaceae)
Hypoxis decumbens (Hypoxidaceae)
Sisyrinchium micranthum (Iridaceae)
Stromanthe stromanthoides (Marantaceae)
Andropogon bicornis (Poaceae)
Axonopus scoparius (Poaceae)
Lasiacis nigra (Poaceae)
Oplismenus hirtellus (Poaceae)
Paspalum saccharoides (Poaceae)
Pennisetum purpureum (Poaceae)
Renealmia sessilifolia (Zingiberaceae)

Appendix B. Dominant or abundant taxa of primary cloud forest (lower montane forest), 1700–2000 m

Trees or arborescent shrubs:
Alsophila erinacea (Cyatheaceae)
Cyathea spp. (Cyatheaceae)
Matisia cf. *malacocalyx* (Bombacaceae)
Brunellia comocladifolia (Brunelliaceae)
Clusia spp. (Clusiaceae)
Weinmannia spp. (Cunoniaceae)
Billia columbiana (Hippocastanaceae)
Nectandra spp. (Lauraceae)
Ocotea spp. (Lauraceae)
Talauma cf. *gilbertoi* (Magnoliaceae)
Blakea spp. (Melastomataceae)
Meriania spp. (Melastomataceae)
Guarea kunthiana (Meliaceae)
Ruagea tomentosa (Meliaceae)
Ficus spp. (Moraceae)
Morus insignis (Moraceae)
Otoba gordoniifolia (Myristicaceae)
Myrcia spp. (Myrtaceae)
Panopsis spp. (Proteaceae)
Faramea spp. (Rubiaceae)
Palicourea amethystina (Rubiaceae)
Posoqueria latifolia (Rubiaceae)
Meliosma spp. (Sabiaceae)
Pouteria cf. *collina* (Sapotaceae)
Brugmansia candida (Solanaceae)
Solanum cucullatum (Solanaceae)
Styrax argenteus (Styracaceae)
Aegiphila alba (Verbenaceae)
Prestoea acuminata (Arecaceae)

Lianas and terrestrial vines:
Mendoncia cf. *orbicularis* (Acanthaceae)
Hebeclinium killipii (Asteraceae)
Mikania spp. (Asteraceae)

Munnozia senecionidis (Asteraceae)
Pentacalia spp. (Asteraceae)
Tourrettia lappacea (Bignoniaceae)
Gurania macrophylla (Cucurbitaceae)
Psiguria cf. *triphylla* (Cucurbitaceae)
Phaseolus polyanthus (Fabaceae)
Hydrangea tarapotensis (Hydrangeaceae)
Stigmaphyllon bogotense (Malpighiaceae)
Marcgravia spp. (Marcgraviaceae)
Odontocarya tripetala (Menispermaceae)
Passiflora spp. (Passifloraceae)
Clematis haenkeana (Ranunculaceae)
Rubus spp. (Rosaceae)
Manettia spp. (Rubiaceae)
Juanulloa pavonii (Solanaceae)
Solanum juglandifolium (Solanaceae)
Trianaea nobilis (Solanaceae)
Tropaeolum spp. (Tropaeolaceae)
Bomarea spp. (Alstroemeriaceae)
Anthurium spp. (Araceae)
Philodendron spp. (Araceae)
Sphaeradenia hamata (Cyclanthaceae)
Dioscorea spp. (Dioscoreaceae)
Smilax spp. (Smilacaceae)

Understory subshrubs, herbs, and pteridophytes:
Huperzia wilsonii (Lycopodiaceae)
Selaginella sericea (Selaginellaceae)
Diplazium spp. (Dryopteridaceae)
Elaphoglossum bakeri (Dryopteridaceae)
Sticherus tomentosus (Gleicheniaceae)
Lophosoria quadripinnata (Lophosoriaceae)
Danaea erecta (Marattiaceae)
Ophioglossum reticulatum (Ophioglossaceae)

Adiantum macrophyllum (Pteridaceae)
Pityrogramma ebenea (Pteridaceae)
Pteris spp. (Pteridaceae)
Thelypteris spp. (Thelypteridaceae)
Clibadium spp. (Asteraceae)
Burmeistera spp. (Campanulaceae)
Centropogon spp. (Campanulaceae)
Podandrogyne sp. (Capparidaceae)
Carica spp. (Caricaceae)
Psammisia spp. (Ericaceae)
Macrocarpaea sodiroana (Gentianaceae)
Alloplectus spp. (Gesneriaceae)
Gasteranthus spp. (Gesneriaceae)
Kohleria spp. (Gesneriaceae)
Ardisia websteri (Myrsinaceae)
Fuchsia macrostigma (Onagraceae)
Piper spp. (Piperaceae)
Galium hypocarpium (Rubiaceae)
Gonzalagunia dependens (Rubiaceae)
Hoffmannia spp. (Rubiaceae)
Nertera granadensis (Rubiaceae)
Psychotria macrophylla (Rubiaceae)
Browallia speciosa (Solanaceae)
Capsicum lycianthoides (Solanaceae)
Pilea spp. (Urticaceae)
Anthurium spp. (Araceae)
Xanthosoma undipes (Araceae)
Costus laevis (Costaceae)
Carex spp. (Cyperaceae)
Heliconia spp. (Heliconiaceae)
Calathea ischnosiphonoides (Marantaceae)
Elleanthus spp. (Orchidaceae)
Erythrodes spp. (Orchidaceae)
Sobralia klotzscheana (Orchidaceae)
Renealmia aurantifera (Zingiberaceae)

Epiphytes and hemiepiphytes:
Huperzia spp. (Lycopodiaceae)
Asplenium spp. (Aspleniaceae)
Blechnum spp. (Blechnaceae)
Elaphoglossum spp. (Dryopteridaceae)

Nephrolepis spp. (Dryopteridaceae)
Melpomene spp. (Grammitidaceae)
Terpsichore spp. (Grammitidaceae)
Hymenophyllum spp. (Hymenophyllaceae)
Trichomanes spp. (Hymenophyllaceae)
Campyloneurum ophiocaulon (Polypodiaceae)
Microgramma spp. (Polypodiaceae)
Niphidium crassifolium (Polypodiaceae)
Pecluma spp. (Polypodiaceae)
Radiovittaria gardneriana (Vittariaceae)
Begonia spp. (Begoniaceae)
Macleania spp. (Ericaceae)
Psammisia spp. (Ericaceae)
Alloplectus spp. (Gesneriaceae)
Columnea spp. (Gesneriaceae)
Drymonia ecuadorensis (Gesneriaceae)
Peperomia spp. (Piperaceae)
Anthurium spp. (Araceae)
Philodendron spp. (Araceae)
Guzmania spp. (Bromeliaceae)
Pitcairnia nigra (Bromeliaceae)
Tillandsia spp. (Bromeliaceae)
Brassavola sp. (Orchidaceae)
Dichaea spp. (Orchidaceae)
Elleanthus spp. (Orchidaceae)
Encyclia vespa (Orchidaceae)
Epidendrum spp. (Orchidaceae)
Lepanthes spp. (Orchidaceae)
Maxillaria spp. (Orchidaceae)
Myoxanthus sp. (Orchidaceae)
Odontoglossum spp. (Orchidaceae)
Oncidium serratum (Orchidaceae)
Pleurothallis spp. (Orchidaceae)
Rodriguezia lehmannii (Orchidaceae)
Stelis spp. (Orchidaceae)
Trichopilia spp. (Orchidaceae)
Xylobium spp. (Orchidaceae)

Appendix C. Dominant or abundant taxa of upper montane forest, 2100–2750 m

Trees and arborescent shrubs:
Cyathea caracasana (Cyatheaceae)
Oreopanax spp. (Araliaceae)
Dendrophorbium lloense (Asteraceae)
Viburnum pichinchense (Caprifoliaceae)
Maytenus macrocarpa (Celastraceae)
Hedyosmum goudotianum (Chloranthaceae)
Clethra obovata (Clethraceae)
Cornus peruviana (Cornaceae)
Sapium stylare (Euphorbiaceae)
Casearia aff. *prunifolia* (Flacourtiaceae)
Citronella incarum (Icacinaceae)
Nectandra obtusata (Lauraceae)
Ocotea spp. (Lauraceae)
Miconia hymenanthera (Melastomataceae)
Miconia theaezans (Melastomataceae)
Siparuna spp. (Monimiaceae)
Stylogyne ambigua (Myrsinaceae)
Eugenia spp. (Myrtaceae)
Myrcianthes discolor (Myrtaceae)
Palicourea calothyrsus (Rubiaceae)
Freziera canescens (Theaceae)

Lianas and terrestrial vines:
Munnozia senecionidis (Asteraceae)
Mutisia grandiflora (Asteraceae)
Oligactis pichinchensis (Asteraceae)
Pentacalia luteynorum (Asteraceae)
Apodanthera sp. (Cucurbitaceae)
Marcgravia spp. (Marcgraviaceae)
Colignonia rufopilosa (Nyctaginaceae)
Passiflora chelidonea (Passifloraceae)
Rubus nubigenus (Rosaceae)
Tropaeolum stipulatum (Tropaeolaceae)
Cissus spp. (Vitaceae)

Bomarea multiflora (Alstroemeriaceae)
Anthurium mindense (Araceae)
Dioscorea spp. (Dioscoreaceae)
Chusquea spp. (Poaceae)

Understory subshrubs, herbs, and pteridophytes:
Huperzia hippuridea (Lycopodiaceae)
Lycopodium spp. (Lycopodiaceae)
Hydrocotyle humboldtii (Apiaceae)
Baccharis latifolia (Asteraceae)
Scybalium depressum (Balanophoraceae)
Burmeistera crispiloba (Campanulaceae)
Centropogon cf. *ferrugineus* (Campanulaceae)
Centropogon nigricans (Campanulaceae)
Coriaria ruscifolia (Coriariaceae)
Weinmannia spp. (Cunoniaceae)
Disterigma acuminatum (Ericaceae)
Gaultheria spp. (Ericaceae)
Pernettya prostrata (Ericaceae)
Escallonia myrtilloides (Escalloniaceae)
Heppiella ulmifolia (Gesneriaceae)
Gunnera atropurpurea (Gunneraceae)
Minthostachys mollis (Lamiaceae)
Fuchsia spp. (Onagraceae)
Thalictrum podocarpum (Ranunculaceae)
Lachemilla spp. (Rosaceae)
Galium hypocarpium (Rubiaceae)
Bartsia mutica (Scrophulariaceae)
Calceolaria crenata (Scrophulariaceae)
Lamourouxia virgata (Scrophulariaceae)
Phenax hirtus (Urticaceae)
Pilea spp. (Urticaceae)
Viola scandens (Violaceae)
Pitcairnia sodiroi (Bromeliaceae)

Uncinia hamata (Cyperaceae)
Epidendrum cochlidium (Orchidaceae)
Chusquea spp. (Poaceae)

Epiphytes and hemiepiphytes:
Blechnum lherminieri (Blechnaceae)
Hymenophyllum jamesonii
(Hymenophyllaceae)
Hymenophyllum myriocarpum
(Hymenophyllaceae)
Trichomanes capillaceum
(Hymenophyllaceae)
Polypodium monosorum (Polypodiaceae)
Begonia truncicola (Begoniaceae)
Macleania macrantha (Ericaceae)
Psammisia sodiroi (Ericaceae)
Sphyrospermum grandifolium (Ericaceae)
Themistoclesia dependens (Ericaceae)
Columnea strigosa (Gesneriaceae)
Peperomia tropaeolifolia (Piperaceae)
Anthurium jimeniae (Araceae)
Anthurium striatipes (Araceae)
Pitcairnia fusca (Bromeliaceae)
Pitcairnia sodiroi (Bromeliaceae)
Racinaea elegans (Bromeliaceae)
Racinaea tetrantha (Bromeliaceae)
Sphaeradenia horrida (Cyclanthaceae)
Brachionidium sp. (Orchidaceae)
Brachtia andina (Orchidaceae)
Epidendrum brevilabre (Orchidaceae)
Masdevallia cf. *ventricularia* (Orchidaceae)
Odontoglossum denticulatum (Orchidaceae)
Oncidium hartwegii (Orchidaceae)
Oncidium pentadactylon (Orchidaceae)
Scaphosepalum sp. (Orchidaceae)
Sertifera sp. (Orchidaceae)

Appendix D. Dominant and conspicuous taxa in riparian vegetation, 1125–2000 m

Equisetum bogotense (Equisetaceae)
Equisetum giganteum (Equisetaceae)
Didymochlaena truncatula (Dryopteridaceae)
Lastreopsis effusa (Dryopteridaceae)
Tectaria antioquiana (Dryopteridaceae)
Sticherus bifidus (Gleicheniaceae)
Adiantum concinnum (Pteridaceae)
Pteris podophylla (Pteridaceae)
Thelypteris spp. (Thelypteridaceae)
Saurauia cf. *prainiana* (Actinidiaceae)
Baccharis nitida (Asteraceae)
Clibadium spp. (Asteraceae)
Erato polymnioides (Asteraceae)
Smallanthus riparius (Asteraceae)
Tessaria integrifolia (Asteraceae)
Begonia foliosa (Begoniaceae)
Alnus acuminata (Betulaceae)
Cecropia monostachya (Cecropiaceae)
Hedyosmum racemosum (Chloranthaceae)
Clusia spp. (Clusiaceae)
Acalypha platyphylla (Euphorbiaceae)
Alchornea sodiroi (Euphorbiaceae)
Phyllanthus sponiifolius (Euphorbiaceae)
Erythrina megistophylla (Fabaceae)
Inga densiflora (Fabaceae)
Mucuna cf. *elliptica* (Fabaceae)
Alloplectus spp. (Gesneriaceae)
Columnea spp. (Gesneriaceae)
Hydrangea peruviana (Hydrangeaceae)
Stachys spp. (Lamiaceae)
Sarcopera anomala (Marcgraviaceae)
Aciotis alata (Melastomataceae)
Arthrostema ciliatum (Melastomataceae)
Blakea spp. (Melastomataceae)
Conostegia superba (Melastomataceae)

Siparuna aspera (Monimiaceae)
Siparuna lepidota (Monimiaceae)
Ficus spp. (Moraceae)
Otoba gordoniifolia (Myristicaceae)
Ardisia sp. (Myrsinaceae)
Ludwigia octovalvis (Onagraceae)
Oxalis psoraleoides (Oxalidaceae)
Piper spp. (Piperaceae)
Rubus bolivianus (Rosaceae)
Arachnothrix reflexa (Rubiaceae)
Hoffmannia spp. (Rubiaceae)
Psychotria macrophylla (Rubiaceae)
Allophylus floribundus (Sapindaceae)
Leucocarpus perfoliatus (Scrophulariaceae)
Capsicum lycianthoides (Solanaceae)
Cuatresia spp. (Solanaceae)
Cyphomandra hypomalaca (Solanaceae)
Urera baccifera (Urticaceae)
Aegiphila albida (Verbenaceae)
Cornutia microcalycina (Verbenaceae)
Anthurium margaricarpum (Araceae)
Anthurium mindense (Araceae)
Anthurium versicolor (Araceae)
Philodendron oligospermum (Araceae)
Canna jaegeriana (Cannaceae)
Tradescantia zanonia (Commelinaceae)
Costus laevis (Costaceae)
Cyperus hermaphroditus (Cyperaceae)
Eleocharis elegans (Cyperaceae)
Eleocharis retroflexa (Cyperaceae)
Xiphidium caeruleum (Haemodoraceae)
Heliconia griggsiana (Heliconiaceae)
Heliconia sclerotricha (Heliconiaceae)
Juncus tenuis (Juncaceae)
Stromanthe stromanthoides (Marantaceae)

Sobralia spp. (Orchidaceae)
Guadua angustifolia (Poaceae)
Gynerium sagittatum (Poaceae)
Heteranthera reniformis (Pontederiaceae)
Typha domingensis (Typhaceae)
Renealmia thyrsoidea (Zingiberaceae)

Appendix E. Distribution of epiphytes in the Maquipucuna vascular flora

Note: Definitions follow Kress (1989) and Williams-Linera & Lawton (1995). Numbers indicate epiphytic species and total number of species, respectively. Facultative species (sometimes terrestrial) are counted as epiphytic.

Holoepiphytes (361)

PTERIDOPHYTES (106)
Lycopodiaceae
  Huperzia (6/12)
Aspleniaceae
  Asplenium (13/19)
Blechnaceae
  Blechnum (5/9)
Dennstedtiaceae
  Dennstaedtia (1/3)
Dryopteridaceae
  Diplazium (3/16)
  Elaphoglossum (16/18)
  Nephrolepis (2/3)
  Polybotrya (2/2)
Grammitidaceae
  Cochlidium (1/1)
  Enterosora (1/1)
  Lellingeria (1/1)
  Melpomene (5/5)
  Terpsichore (6/6)
Hymenophyllaceae
  Hymenophyllum (11/11)
  Trichomanes (3/5)
Polypodiaceae
  Campyloneurum (9/9)
  Microgramma (3/3)
  Niphidium (1/1)
  Pecluma (3/3)

Polypodium (10?/14)
Thelypteridaceae
  Thelypteris (1/15)
Vittariaceae
  Polytaenium (1/1)
  Radiovittaria (2/2)

DICOTS (69)
Begoniaceae
  Begonia (6/15)
Campanulaceae
  Burmeistera (1/5)
Ericaceae
  Cavendishia (1/3)
  Disterigma (1/2)
  Macleania (4/6)
  Psammisia (3/5)
  Sphyrospermum (3/4)
  Themistoclesia (1/1)
Gesneriaceae
  Alloplectus (6/8)
  Columnea (11?/11)
  Drymonia (1/3)
Loranthaceae
  Dendrophthora (1/1)
  Oryctanthus (1/1)
  Phoradendron (4/4)
  Struthanthus (2/2)
Piperaceae
  Peperomia (17?/41)

Piper (2/38)
Rubiaceae
  Hillia (1/1)
Urticaceae
  Pilea (3/14)

MONOCOTS (186)
Araceae
  Anthurium (9?/40)
Bromeliaceae
  Catopsis (1/1)
  Guzmania (14/16)
  Mezobromelia (2/2)
  Pitcairnia (2/7)
  Racinaea (4/4)
  Tillandsia (4/5)
Orchidaceae
  Ada (2/2)
  Barbosella (1/1)
  Brachtia (1/1)
  Brassavola (1/1)
  Brassia (2/2)
  Chondrorhyncha (2/2)
  Comparettia (1/1)
  Cryptocentrum (1/1)
  Dichaea (5/5)
  Dracula (3/3)
  Dryadella (1/1)
  Elleanthus (6?/12)
  Encyclia (4/4)

Epidendrum (22?/31)
Eurystyles (1/1)
Isochilus (1/1)
Kefersteinia (2/2)
Lepanthes (8?/8)
Lockhartia (1/1)
Lycaste (1?/2)
Lycomormium (1/1)
Macroclinium (1?/1)
Masdevallia (8?/8)
Maxillaria (14?/14)
Myoxanthus (2/2)
Myrosmodes (1/1)
Odontoglossum (3/4)
Oncidium (8?/8)
Peristeria (1/1)
Platystele (1/1)
Pleurothallis (13/13)
Polystachya (1/1)
Porroglossum (3/3)
Psygmorchis (1/1)
Restrepiopsis (1/1)
Rodriguezia (1/1)
Scaphosepalum (1/1)
Scelochilus (3/3)
Sertifera (1?/2)
Sigmatostalix (1/1)
Stanhopea (2/2)
Stelis (3/3)
Stellilabium (3/3)
Symphyglossum (1/1)
Trichopilia (2/2)
Trichosalpinx (2/2)
Trisetella (1/1)
Xylobium (3/3)
Zootrophion (1/1)

Hemiepiphytes (80)

(Primary)

DICOTS (25)
 Araliaceae
  Oreopanax (1/5)
  Schefflera (5?/5)
 Bignoniaceae
  Schlegelia (1/2)
 Clusiaceae
  Chrysochlamys (2?/2)
  Clusia (7?/7)
 Hydrangeaceae
  Hydrangea (2/2)
 Melastomataceae
  Blakea (2?/4)
 Moraceae
  Ficus (1/13)
 Solanaceae
  Juanulloa (1?/1)
  Markea (1/1)
  Schultesianthus (1/1)
  Trianaea (1/1)

(Secondary)

DICOTS (7)
 Marcgraviaceae
  Marcgravia (5?/5)
  Sarcopera (2?/2)

MONOCOTS (48)
 Araceae
  Anthurium (25/40)
  Monstera (2/2)
  Philodendron (15?/16)
 Cyclanthaceae
  Asplundia (3/4?)
  Ludovia (1/1)
  Sphaeradenia (2/2)

Literature Cited

Acosta-Solís, M. 1966. Las divisiones fitogeográficas y las formaciones geobotánicas del Ecuador. Revista Acad. Colomb. Ci. Exact. 12: 401–447.

Acosta-Solís, M. 1968. Naturalistas y viajeros científicos que han contribuido al conocimiento florístico y fitogeográfico del Ecuador. Contr. Inst. Ecuat. Ci. Nat. 65: 1–138.

Balslev, H. 1988. Distribution patterns of Ecuadorean plant species. Taxon 37: 567–577.

Balslev, H., & S. Renner. 1989. The diversity of the Ecuadorean forests east of the Andes. *In* L. B. Holm-Nielsen, I. Nielsen, & H. Balslev (eds.), Tropical Forests: Botanical Dynamics, Speciation and Diversity, p. 287–295. London: Academic Press.

Blandín Landívar, C. 1977. El Clima y sus Características en el Ecuador. Quito: Biblioteca Ecuador.

Cerón M., C. 1993. Etnobotánica del Ecuador: estudios regionales. Quito: Ediciones Abya-Yala.

Cerón M., C., & L. Ávila. 1994. Anexo 4, Diagnósticos botánicas. *In* J. Calvopiña, X. Izurieta, R. Monoselvas, & R. Ulloa, Plan de manejo del bosque y vegetación "Montañas de Mindo y Cordillera de Nambillo", p. 35–51. Quito: Ecociencia y Corporación Ecológica "Amigos de la Naturaleza Mindo."

Churchill, S. P., H. Balslev, E. Forero, & J. L. Luteyn (eds.) 1995. Biodiversity and Conservation of Neotropical Montane Forests. Bronx NY: New York Botanical Garden.

Cuamacás, S. B., & G. A. Tipaz. 1995. Árboles de los Bosques Interandinos del Norte del Ecuador. Quito: Publ. Museo Ecuatoriano de Ciencias Naturales, Monografia 4.

Diels, L. 1937. Beiträge zur Kenntnis der Vegetation und Flora von Ecuador. Biblioth. Bot. 116: 1–175.

Dillon, M. O., A. S. Alva, I. S. Vega, S. Ll. Quiroz, & N. Hensold. 1995. Floristic inventory and biogeographic analysis of montane forests in northwestern Peru. *In* S. P. Churchill et al. (eds.), Biodiversity and Conservation of Neotropical Montane Forests, p. 251–269. Bronx NY: New York Botanical Garden.

Dodson, C. H., & R. Escobar R. 1996. Orquídeas Nativas del Ecuador, vol. I: *Aa–Dracula*. Medellin: Editorial Colina.

Dodson, C. H., & A. H. Gentry. 1978. Flora of the Río Palenque Science Center. Selbyana 4: 1–628.

Dodson, C. H., & A. H. Gentry. 1993. Extinción biológica en el Ecuador occidental. *In* P. A. Mena & L. Suárez (eds.), La Investigación para la Conservación de la Diversidad Biológica en el Ecuador, p. 27–57. Quito: Ecociencia.

Engler, A. 1905. Araceae-Pothoideae. Das Pflanzenreich IV. 23B (Heft 21): 1–330.

Estrella, E. (ed.). 1989. [Introduction to] Flora Huayaquilensis... auctore Johanne Tafalla. Tomus I. Madrid: Editio Facta AB.

Forero, E., & A. H. Gentry. 1989. Lista Anotada de las Plantas del Departamento del Chocó, Colombia. Bibliot. José Jerónimo Triana 10: 1–142.

Foster, R. B., M. R. Metz, & G. L. Webster. 1999. Plantas Llamativas y Comunes de la Reserva Maquipucuna, Pichincha, Ecuador. Chicago: Environmental and Conservation Programs, Field Museum.

Gentry, A. H. 1986. Species richness and floristic composition of Chocó region plant communities. Caldasia 15: 71–91.

Gentry, A. H. 1990. Floristic similarities and differences between southern Central America and upper and central Amazonia. *In* A. H. Gentry (ed.), Four Neotropical Rainforests, p. 141–157. New Haven: Yale University Press.

Gentry, A. H. 1992. Diversity and floristic composition of Andean forests of Peru and adjacent countries: implications for their conservation. Mem. Mus. Hist. Nat. "Javier Prado" 21: 11–29.

Gentry, A. H. 1993. A Field Guide to the Families and Genera of Woody Plants of Northwest South America. Washington DC: Conservation International.

Gentry, A. H. 1995. Patterns of diversity and floristic composition in Neotropical montane forests. *In* S. P. Churchill et al. (eds.), Biodiversity and Conservation of Neotropical Montane Forests, p. 103–126. Bronx NY: New York Botanical Garden.

Greenfield, P. 1993. Bird-list of the Maquipucuna Reserve and Bosque Protector. Unpublished manuscript.

Grubb, P. J. 1977. Control of forest growth and distribution on wet tropical mountains: with special reference to mineral nutrition. Annual Rev. Ecol. Syst. 8: 83–107.

Haber, W. A. 2000. Plants and vegetation; Appendix 1: Vascular plants of Monteverde. *In* N. M. Nadkarni and N. T. Wheelwright (eds.), Monteverde: Ecology and Conservation of a Tropical Cloud Forest. New York: Oxford University Press.

Hammel, B. 1990. The distribution of diversity among families, genera, and habit types in the La Selva flora. *In* A. H. Gentry (ed.), Four Neotropical Rainforests. p. 75–84. New Haven: Yale University Press.

Humboldt, A. von. 1807. Ideen zur einer Geographie der Pflanzen [reprint 1963]. Darmstadt: Wissenschaftlichen Buchgesellschaft.

Jaramillo, J. L., & E. Grijalva. 1993. Inventario florístico preliminar y manejo de especies nativas de la Reserva Florística Río Guajalito. *In* R. Valencia & K. Romoleroux (eds.), Trabajos Ecuatorianos en Botánica: Primer Congreso Ecuatoriano de Botánica. Quito: FUNBOTANICA.

Jørgensen, P. M., & S. León-Yánez. (eds.) 1999. Catalogue of the Vascular Plants of Ecuador. St. Louis: Missouri Botanical Garden Press.

Jørgensen, P. M., & C. Ulloa U. 1989. Estudios Botánicos en la "Reserva ENDESA", Pichincha, Ecuador. AAU Rep. 22: 1–138.

Jørgensen, P. M., & C. Ulloa U. 1994. Seed plants of the High Andes of Ecuador—a checklist. AAU Rep. 34: 1–443.

Jørgensen, P. M., & R. Valencia. 1988. Composición y estructura de un bosque andino: Pasochoa, Ecuador. Publ. Mus. Ecuat. Ci. Nat. 8: 21–38.

Justicia, R. 1995. The Chocó-Andean Biological Corridor (CABCOR). Unpublished manuscript.

Kessler, M. 1995. Biogeography, Vegetation. *In* B. J. Best & M. Kessler, Biodiversity and Conservation in Tumbesian Ecuador and Peru, p. 41–118. Cambridge: BirdLife International.

Kress, W. J. 1989. The systematic distribution of vascular epiphytes. *In* U. Lüttge (ed.), Vascular Plants as Epiphytes, p. 234–261. Berlin: Springer-Verlag.

Lauer, W. 1986. Die Vegetationszonierung der Neotropis und ihr Wandel seit der Eiszeit. Ber. Deutsch. Bot. Ges. 99: 211–235.

Lima, M. P. M. de, & R. R. Guedes-Bruni. 1994. Reserva Ecológica de Macaé de Cima (Novo Friburgo–RJ): Aspectos florísticos das espécies vasculares, vol. 1. Rio de Janeiro: Jardim Botánico.

Mansour, J. (ed.). 1995. Parks in Peril Source Book. Washington: The Nature Conservancy.

Marín, M., et al. 1992. The Maquipucuna Reserve bird list. Unpublished manuscript.

Meier, W. 1998. Flora und Vegetation des Ávila-Nationalparks (Venezuela/ Küstenkordillere) unter besonderer Berücksichtigung der Nebelwaldstufe. Dissertationes Botanicae 296: 1–485. Berlin: J. Cramer.

Mena Vásconez, P. 1995. Las áreas protegidas con bosque montano en Ecuador. *In* S. P. Churchill et al. (eds.), Biodiversity and Conservation of Neotropical Montane Forests, p. 627–635. Bronx NY: New York Botanical Garden.

Neill, D. A., & B. Øllgaard. 1993. Los inventarios botánicas en el Ecuador: estado actual y prioridades. *In* P. A. Mena & L. Suárez, La Investigación para la Conservación de la Diversidad Biológica en el Ecuador, p. 61–80. Quito: Ecociencia.

Nicolson, D. H. 1983. Sodiro's publications on Araceae. Huntia 5: 3–15.

Padilla, V. 1983. Foreword. *In* E. F. André, Bromeliaceae Andreanae (transl. M. Rothenberg). Berkeley: Two Windows Press.

Phillips, O. L., & J. Miller. "1996". Global Patterns of Plant Diversity: Alwyn H. Gentry's Forest Transect Data Set. Monogr. Syst. Bot. Missouri Bot. Gard. (ined.?). [Cited by Phillips & Raven (1996), but apparently still unpublished.]

Phillips, O. L., & P. H. Raven. 1996. A strategy for sampling neotropical forests. *In* Gibson, A. C. (ed.), Neotropical Biodiversity and Conservation. Occ. Publ. Mildred E. Mathias Bot. Gard. 1: 141–155. Los Angeles: University of California.

Pipoly, J. 1996. New species of *Ardisia* (Myrsinaceae) from Ecuador and Peru. Sida 17: 445–458.

Raguso, R. A. 1996. Preliminary checklist and field observations of the butterflies of the Maquipucuna field station (Pichincha Province, Ecuador). J. Res. Lepid. 32: 135–161.

Renner, S. S. 1993. A history of botanical exploration in Amazonian Ecuador, 1739–1988. Smithsonian Contr. Bot. 82: 1–39.

Reyes, E. A.1993. Los Yumbos de Rumicucho. Quito: Ediciones Abya-Yala.

Rios, M. 1993. Plantas útiles en el noroccidente de la provincia de Pichincha. Quito: Ediciones Abya-Yala.

Robinson, H. 1997. New species of *Aphanactis, Calea, Clibadium*, and *Tridax* (Heliantheae: Asteraceae) from Ecuador and Peru. Phytologia 82: 58–62.

Saloman, F. 1997. Los Yumbos, Niguas y Tsatchila o "Colorados" durante la colonia española. Quito: Ediciones Abya-Yala.

Sarmiento, F. O. 1994. Human impacts on cloud forests of the upper Guayllabamba River basin, Ecuador, and suggested management responses. *In* L. S. Hamilton, J. O. Juvik, & F. N. Scatena (eds.), Tropical Montane Cloud Forests, p. 284–295. Berlin: Springer-Verlag.

Sarmiento, F. O. 1995. Restoration of equatorial Andes: the challenge for conservation of Trop-Andean landscapes in Ecuador. *In* S. P. Churchill et al. (eds.), Biodiversity and Conservation of Neotropical Montane Forests, p. 637–651. Bronx NY: New York Botanical Garden.

Silverstone-Sopkin, P. A., & J. E. Ramos-Pérez. 1995. Floristic exploration and phytogeography of the Cerro del Torrá, Chocó, Colombia. *In* S. P. Churchill et al. (eds.), Biodiversity and Conservation of Neotropical Montane Forests, p. 169–186. Bronx NY: New York Botanical Garden.

Sodiro, L. 1903. Antúrios Ecuatorianos. Contribuciones al Conocimiento de la Flora Ecuatoriana, Monografia II. Quito: Escuela de Artes y Oficios.

Steyermark, J. A., & O. Huber. 1978. Flora del Ávila. Caracas: Sociedad Venezolana de Ciencias Naturales & Ministerio del Ambiente y de los Recursos Naturales Renovables.

Svenning, J.-C., & H. Balslev. 1998. The palm flora of the Maquipucuna montane forest reserve, Ecuador. Principes 42: 218–226.

Terán, F. 1984. Geografía del Ecuador. Quito: Libresa.

Ulloa, C. U., & P. M. Jørgensen. 1993. Árboles y arbustos de los Andes del Ecuador. AAU Rep. 30: 1–264.

Valencia R., R. 1995. Composition and structure of an Andean forest fragment in eastern Ecuador. *In* S. P. Churchill et al. (eds.), Biodiversity and Conservation of Neotropical Montane Forests, p. 239–249. Bronx NY: New York Botanical Garden.

Vázquez-García, J. A. 1995. Cloud forest archipelagos: preservation of fragmented montane ecosystems in tropical America. *In* L. S. Hamilton, J. O. Juvik, & F. N. Scatena (eds.), Tropical Montane Cloud Forests. Berlin: Springer-Verlag.

Webster, G. L. 1995. The panorama of Neotropical cloud forests. *In* S. P. Churchill et al. (eds.), Biodiversity and Conservation of Neotropical Montane Forests, p. 53–77. Bronx NY: New York Botanical Garden.

Williams-Linera, G., & R. O. Lawton. 1995. The ecology of hemiepiphytes in forest canopies. *In* Lowman, M. D., & N. N. Nadkarni (eds.), Forest Canopies, p. 255–283. San Diego: Academic Press.

Young, K. R. 1991. Floristic diversity on the eastern slopes of the Peruvian Andes. Candollea 46: 125–143.

Zahawi, R. A., & C. K. Augspurger. 1999. Early plant succession in abandoned pastures in Ecuador. Biotropica 31: 540–552.

Plates

All photographs are by Grady L. Webster except for
Plates 8A, 8B, 10B, 13B, and 15B, which are by Jim Cooper.

Plate 1: University Research Expeditions Program (UREP) participants, 1989-1995; A, 1989; B, 1990; C, 1992; D, 1993; E, 1995.

Plate 2: A, Original Maquipucuna "Hilton", 1989;
B, New Maquipucuna "Hilton", 1995.

Plate 3: A, Steep slopes along road from Calacalí to Nanegalito; B, Riparian forest along Río Umachaca.

Plate 4: A, View of Nanegalito from Cerro Negro; B, Cloud forest on steep ridges along Río Pichán (streambed at 1900 m).

Plate 5: A, Primary cloud forest on main ridge of Cerro Sosa (Montañas de Maquipucuna), near Base Camp 1, c. 1750 m; B, View from main ridge of Cerro Sosa at c. 1500 m, disturbed cloud forest.

Plate 6: A, *Burmeistera multiflora* (Campanulaceae);
B, *Psiguria* cf. *triphylla* (Cucurbitaceae).

Plate 7: A, *Cecropia monostachya* (Cecropiaceae); B, *Caverdishia tarapotana* (Ericaceae).

Plate 8: A, *Alloplectus bolivianus* (Gesneriaceae); B, *Columnea picta* (Gesneriaceae).

Plate 9: A, *Gunnera brephogea* (Gunneraceae); B, *Ardisia websteri* (Myrsinaceae).

Plate 10: A, *Meriana maxima* (Melastomataceae); B, *Monnina pseudopilosa* (Polygalaceae).

Plate 11: *Ladenbergia pavonii* (Rubiaceae).

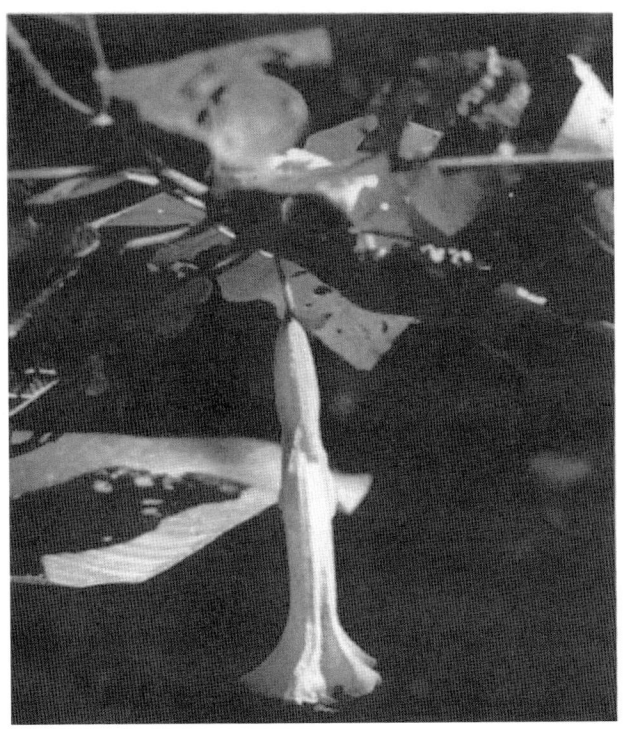

Plate 12: A, *Brugmansia candida* (Solanaceae);
B, *Solanum juglandifolium* (Solanaceae).

Plate 13: A, *Furcraea* cf. *andina* (Agavaceae); B, *Bomarea pardina* (Alstroemeriaceae).

Plate 14: A, *Eucrosia dodsonii* (Amaryllidaceae); B, *Philodendron acuminatissimum* (Araceae).

Plate 15: A, *Geonoma undata* (Arecaceae); B, *Guzmania wittmackii* (Bromeliaceae).

Plate 16: A, *Heliconia impudica* (Heliconiaceae); B, *Epidendrum* cf. *tungurahuae* (Orchidaceae).

Plate 17: *Odontoglossum cirrhosum* (Orchidaceae).